Inspiring Palita Perspectives

Preface

'Palita' means old or grey-haired or ancient. This book is an attempt to share some interesting perspectives from ancient Indian scriptures as I read and understand them. I find these perspectives rational and thought provoking.

Chandogya Upanishad says pRccha, panditaM and medhavat are the ways a person learns. pRccha is questioning. PanditaM is investigation. Medhavat is self-judgement or decision making. I dedicate this collection of essays to these qualities called for by the 'palita' perspectives.

Introduction

The common theme running through several ancient Indian scriptures can be summed up as follows: They ask us to be detached "observers" or "witnesses" to the actions of ourselves and others. Such an observer is our 'Guru'. The observer is also embedded in the blueprint/design pattern of the Universe's evolution as explained in texts like samkhya kArika.

Is there a design pattern in the Universe's evolution? If so, can we begin to understand/comprehend it? If we can comprehend it, is there anything to be learnt from it? What does its existence mean for the nature of the Universe- is it purely deterministic or is there room for randomness? Empirical observations show us that randomness is manifest in our universe and even in our consciousness. How can we reconcile these seemingly opposite conclusions?

In the essays that follow, I have tried to grapple with some of the above questions. Using a curated subset of texts and shlokas, I hope to shine a light on the overarching principles that I believe bind the disparate elements of our ancient philosophies together. While the subset is not exhaustive, I intend to use them as a filter through which we can view several modern ideas and science, thereby creating syncretic interpretation of both modernity and tradition.

This collection contains texts and relevant translations from several sources like Samkhya kArika, Guru Gita, Ashtavakra gita, Chandogya Upanishad, Vishnu SahasranAma and Brihadaranyaka Upanishad. While I only quote them selectively, these individual sources have generated a vast trove of literature/commentaries by themselves. In my translations and understanding, the complete texts also reinforce the very thoughts presented here in this discussion.

I thank Sriram Venkateswaran for reviewing the book, challenging the thoughts presented herein with a great discussion, rewriting parts of it and thus and helping me to refine the presentation. I also thank TV Sreenivas, Shirish B Purohit, Shanthi Ganesan and Ramya Lakshminarayanan for their reviews and feedback.

The 'gist'

The central thesis explored in this book is the presence of a design pattern in the Universe across several domains, which is highlighted metaphorically in ancient scriptures. Several science concepts are also presented through the lens of this design pattern.

The slokas of ashtavakra gita provide a good starting point for this exploration. The slokas call for developing a witnessing or observing component in our personality. By developing thoughts detached from our body and its actions, the dispassionate witnessing trains the development of an observer of manas (manas sAksi), where manas are the thoughts attached to our body and its actions. Such an observer gives liberation from bondings and facilitates knowledge evolution.

The causal relationship between observation and reality as propounded in Brihadaranyaka Upanishad which seems applicable from quantum physics to the theory of mind and everything in between, is then touched upon. Different individuals perceive different realities because their observations are different. Individual observations form 'maya' (mine), a self-centric perspective or subjectivity. When cloud of subjectivity could not be broken and individuals remain attached to personal perspectives, they inhabit a world blurred by 'mAyA' (illusion). A well-trained observer (manas-sAkshi) enables bringing in multiple perspectives and overcome the 'mAyA'.

Comparing our perceptions to a reflection in a mirror, a stanza from dakSinamurthy stotram which metaphorically stakes out the difference between an objective reality and our clouded perceptions is then explained. The coexistence of these two different entities is named 'dvaita'. The slokas also compare inhabiting a personal reality clouded by mAyA to the act of sleeping. While we are asleep, we dream and do not perceive any form of objective reality. The slokas name and describe the entity that wakes us up and helps us overcome mAyA as Guru, the Atman. Atman or Guru they explain, is the detached, dispassionate witness or observer within us as described in other texts.

The concept of 'Guru' is further expanded with Guru Gita which defines Guru as none other than Buddhi-Atman. Buddhi is the intelligence that drives thoughts and actions. The detached witness is the Atman. When this Atman drives the buddhi (intelligence), that buddhi-Atman is the Guru.

Then the book delves deeper into the mechanics of how the observer facilitates evolution. The fundamental design pattern of the universe contains three essential components - evolution or change, an observer/witness that facilitates evolution but is unaffected by evolution, and an object(s) that is subject to evolution. This design mantra is repeatedly observed across ideas in different domains, ranging from quantum physics to the theory of mind and everything in between. This design mantra is what is described in saMkhya kArika.

Samkhya kArika talks of 24 principles of samkhya, the above design pattern. The observer/witness that facilitate evolution, but not affected by evolution is the puruSa. The object that undergoes evolution is the prakRti. The gunas are the three fundamental characteristics of prakRti that get modified during its evolution. They are Sattva, Rajas and Tamas. The 24 principles of evolution are said to be

- prakRti the object of evolution
- Seven differentiations of prakRiti which are
    - Mahat
    - AhamkAra
    - Pancha-bhutAs
- Sixteen modifiers/vikAras which are

    - Five modifiers of mahat (pancha buddhi indriyas)
    - Five modifiers of ahamkAra (pancha karma indriyas)
    - Five measures of pancha-bhutAs (pancha tanmAtras)
    - Atman or puruSa, which is the ultimate modifier that just remains as an observer.
- The 1+7+16 becomes 24 principles.

The evolution of prakRti manifests through change in the balance of three gunas of prakRti. The three gunas are Sattva, Rajas and Tamas. Evolution begins with Sattva acquiring preponderance over other gunas. Then Rajas and Tamas acquire dominance one after another. On this again Sattva acquires predominance and the cycle continues. This is explained in Chandogya Upanishad Chapter 6.

Chandogya Upanishad Chapter six has the mahAvAkya 'Tat Tvam Asi'. It talks of three forms of modification (gunas) causing all the differentiation in Universal matter and beings. It calls them Teja, Apa and Annam. Teja is the Sattva. Apa is the Rajas. Annam is the Tamas. It talks of the design pattern in tri-gunas across multiple domains with different examples. It says Atman is 'Satyam'. Sat is existence or manifestation and hence equated to truth. Satyam is translated as the property of existence in all the matter and beings that exist. This Atman manifests in biological beings as Observer that guides in the thoughts, as Consciousness in biological beings and even in all the actions of beings surrounded by falsehood or truths.

If Atman is the property of existence in all matter and beings, then how do we perceive 'I'...? This question is addressed next. The perception of 'I' is due to expression of ahamkAra. When the internal energy of a matter or being allows it to not just exist as an independent entity alone but also enables it to build and evolve further, then it is said to have ahamkAra.

Many matter forms and beings have ahamkAra. But more complex forms have more ways to express ahamkAra, their internal property that allows them to build and evolve further. Human beings at the top of the evolutionary chain can express ahamkAra through their thoughts and actions too. This expression of ahamkAra through thoughts causes the perception of 'I'. It also leads to the desire of living long and the attempts to link 'I' with the Atman, which is actually beyond all perception.

If there is a design pattern, is the Universe deterministic or came out of random chaos.? This is addressed next. Quantum and thoughts domain are random in their very manifestation. Though they are random, there is a Rta or predictability or order in the Universe. The laws of the Universe are fixed and are called the 'dharma'. This dharma brings in a predictability or order or Rta in the domain it operates. For example, the laws of Universe such as 'observer effect' collapses the probabilistic wave nature to deterministic particle nature and brings up Rta, the order in the Universe though quantum domain is random. This leads to evolution of the classical matter forms and eventually the biological forms. In the same way our thoughts are random in nature. But observation of these thoughts brings up a Rta or order and facilitate their evolution.

Thus, observer facilitating the evolution is a design pattern across multiple domains. For human beings it translates into developing a personality of being a detached witness to ourselves and our actions, which is what is described in texts like ashtavakra Gita.

Some of the readers might think that the above explanations are a 'fitting the curve' exercise. It is and it is not. The above is not 'fitting the curve' exercise because the texts have been 'translated' and not interpreted. The translations have been given and interested readers can engage in further discussions on the accuracy of translations. The above is a 'fitting the curve' exercise too because the mappings of the design pattern across multiple-domains is not from the scriptures. The readers are best assured that lot of efforts have been taken with all sincerity and honesty not to present any erroneous information either on science or on the ancient scriptures side.

Become an Observer

In Ashtavakra Gita[2], Janaka asks Ashtavakra, how one can obtain jnAna (knowledge/wisdom), achieve muktir (freedom from worries) and attain vairAgyam (become resolute). Ashtavakra explains that characteristics of Ksama (patience), arjva (sincerity or candidness), daya (kindness), toSa (contentment) and Satyam (truthfulness) have to be accepted like amRtam or nectar to obtain jnAna, achieve muktir and attain vairAgyam. All these characteristics arise from our thoughts.

1 MAKE YOUR THOUGHTS A WITNESS

Ashtavakra gita elaborates further in Slokas 1.3 to 1.7. The perception of the self and the other, the "I" and "you" are emergent phenomena that arise at the juncture of interacting thoughts. The attributes that we associate with ourselves - our characteristics arise from our thoughts. These Individual characteristics are expressed by our thoughts which impact other's thoughts. Thus, we inherit characteristics and also acquire/develop characteristics from other's thoughts. Thus, each of 'us' are a thought bundle with our own unique 'thought pattern'.

Ashtavakra Gita tells how one needs to train the thoughts. In general, thoughts are associated with physical body. So, the 'I' develops. Once 'I' develops, there is 'you', 'we', 'us' and others.

Concepts of dharma, adharma, pleasure and pain arise from these associations. Concepts of varna, asrama and sensual pleasures arise from these associations. The 'I' sees the bodies as made of materials of this Universe.

What needs to be done is to 'detach' thoughts from the body and make thoughts as observers of the host body, other bodies, other thoughts and everything in this Universe. Become a witness to 'yourself' and others. In short develop a component of your personality as being an observer of yourself and others.

Atman is said to manifest as puruSa[8], the Observer or Witness (sAkshi) of this Universe. When thoughts become observers or witnesses of this Universe like the puruSa, Atman is said to manifest in us. Then like Atman or the puruSa our thoughts become neither doers nor enjoyers and become detached or free from all pains and pleasures.

2 SUMMARY OF SLOKA 1.3 TO 1.7

The characteristics of Ksama (Patience), arjva (sincerity/candidness), dayA (kindness), toSa (Contentment) and Satyam (Truthfulness) have to be accepted like 'nectar' to acquire mukti (freedom from desires), jnAna (knowledge/wisdom) and vairAgyam (determination).

'You' are neither earth, fire, water, air or ether. To become free, realize your 'thoughts' as the observers/witness like Atman. sAkshi which means witness is derived from sa+akSi, which means that eye. Atman is the 'observing eye' of the Universe.

As long as our thoughts are operating around our physical body, we are bound. Once our thoughts move away from the physical body, our thoughts become free and happy.

When our thoughts are around the physical body, we consider ourselves belonging to varna, asrama, as dwelling place of senses etc. But once our thoughts move away from the physical body, they become unattached, vacant observers of the Universe, ever happy.

Our manas, the thoughts that control our actions, brings in the concepts of dharma, adharma, pleasure and pain, as long as our thoughts operate around the physical body. These concepts are not the greatest (they are not the Lord).

Being an observer of everything, one is liberated of everything. When the seer does see more, then alone the bonds (happen). When thoughts become the observers/witness of our body, observing our body and its actions as that of other bodies and actions in the Universe, we become neither the doers, nor the enjoyers, free from everything.

3 Sloka 1.3 to 1.7

na pṛthvī na jalaṁ nāgnirna vāyurdyaurna vā bhavān

eṣāṁ sākṣiṇamātmānaṁ cidrūpaṁ viddhi muktaye

You (bhavAn) are neither prthvi (na prthvi) nor jalam (na jalam) nor agni (na agni) nor vayu (na vAyu) nor dyaus (na dyaur). For becoming free (muktaye), Know/realize (viddhi) the 'cit-rupam', the 'thought-form' (cid rupam) the witness/observer (sAkshi) of the Atman (Atman).

yadi dehaṁ pṛthak kṛtya citi viśrāmya tiṣṭhasi

adhunaiva sukhī śānto bandhamukto bhaviṣyasi

If/Once body is separated (prthak krtya), thoughts remain/situated at rest/peace (visramya tisthasi). Right then (adhunaiva), they become happy (sukhi santo) and will become free (bhavisyasi) from bondages. (bandhamukta)

na tvaṁ viprādiko varṇo nāśramī nākṣagocaraḥ

asaṅgo'si nirākāro viśvasākṣī sukhī bhava

'You' are neither varna of Vipra etc, nor asrama, nor the dwelling place (gocarah) of your senses (akSa). Unattached, formless/vacant, observer/witness of the universe become happy.

dharmādharmau sukhaṁ duḥkhaṁ mānasāni na te vibho
na kartāsi na bhoktāsi mukta evāsi sarvadā

dharma, adharma, duhkham sukham are derived from the manas (mAnasAni), not (na) those (te) the lord/the great (vibho). Neither the doer, nor the enjoyer, become free from everything.

eko draṣṭāsi sarvasya muktaprāyo'si sarvadā

ayameva hi te bandho draṣṭāraṁ paśyasī taram

The one (eko) observer of (drstasi) everything (sarvasya) is liberated/free (mukta prAya) of everything (sarvada). This alone (ayam eva hi) the bonds (te bandho) the seer (drastaram) do see more (pasyasi taram).

Reality and Illusion

4 sAkSi and cakSu

Atman is said to be the 'sa-aksi' (the 'eye') or witness or observer of the Universe. If our thoughts become observers of our own body and its actions as well as others, then Atman is said to manifest in us.

cakSu is the sight or observation. Observation of the existence of something leads to our 'reality' or 'Satyata'. Since our observations vary, our 'realities' vary. To some extent, the differences in our realities caused by differences in the observation, can be understood, measured and even reconciled.

Brihadaranyaka Upanishad sloka 4.1.4 says 'cakSusa satyata'[1]. Our reality (satyata) is based on our observation. This is a Universal principle that applies to all the domains of our Universe.

5 Observation leads to 'reality'

In Quantum domain, 'cakSusa satyata' is observation leading to wave-function collapse.

In Classical domain, 'cakSusa satyata' is speeds of objects/particles varying with respect to observing frame of reference.

In the biological domain, 'cakSusa satyata' is different human beings perceiving different realities of the same event, due to different perceptions caused by their individual observations.

All of these differences due to observation can be measured and their differences reconciled. But in human beings, measuring this difference in observation and reconciliation is not possible at times, due to a phenomenon called mAyA, which is translated as 'illusion' or 'ignorance.

6 mAyA - The 'self-centric' illusion

'ma' means 'my'. 'maya' is 'myself' or 'mine'. This 'maya' or 'mine' gives rise to a self-centric perspective. mAya is descended from maya, the 'self-centric perspective' or 'illusion' or 'ignorance'.

mAya makes us see what we wish or want to see. Our mind wants us to see things as aligned with what it already knows. This reinforcement of our existing thoughts, our self-perspectives, our self-property is mAyA, that arises out of the 'maya'. mAya is the descendant of maya.

This mAya is the illusion or ignorance that our mind creates, as that's the observation it is comfortable with. Our mind does not like conflicting information which causes pain. Hence it tries to align any information received to what exists already in it. This is the illusion that our mind creates, leading to ignorance of the reality. Hence mAya leads to ignorance.

mAyA leads us to being unable to reconcile our differences with others. Without mAya, we will agree to disagree as we understand others perspectives. With mAyA, we don't allow multiple perspectives or conflicting information to come into us. This in turn, leads to us becoming passionate as we keep reinforcing the same things again and again about what we believe is true. In comparison, if we don't have mAyA, if we allow multiple, conflicting perspectives to come into us, we become more knowledgeable.

7 RECONCILING SATYATA AND MAYA

The true reality is fundamentally limited by our observation. Our observation is limited by our genetic and environmentally acquired characteristics, which are our vAsanas. Everyone has his / her own observations and realities due to their own vAsanas.

Hence our perception of absolute truth is limited by our vAsanas. Hence our realities are different by 'default'. Our realities also vary with time. What is satyata at a point of time amongst a group of people may not be the satyata at a different point of time with the same or another group of people.

Yes, people can align on these observational differences by sharing their perspectives. But his sharing of perspectives is possible only when people move out of mAyA, the illusion caused by self-perspectives. Thus, our ancient texts teach us to listen and understand more of those perspectives that are different from our own. That's the only way we can remove mAyA.

Shift the 'seer' or 'observer' from maya, the self. Move away from maya (self-view) to how others see/perceive. When we are able to visualize perspective of multiple stakeholders, (simply put, thinking from other's shoes), maya the self-perspective and mAya the resultant illusion goes away.

When we visualize views of different stakeholders, from their side, it raises questions in our mind, creates doubts on our 'self' perspective. It may agitate our mind and unsettle it. But when these multiple views are reconciled, the mAyA reduces a lot. We get closer to satyata, the existence or reality.

Wakeup from mAyA

The illusion/ignorance causing self-perspective, mAyA is like a sleep. In dakShinamUrthy stotram[3], Adi Shankaracharya explains the concept of mAya with examples such as a city seen in a mirror and our perception and reality being different while sleeping and becoming same when we wake up.

8 UNIVERSE - CITY SEEN IN A MIRROR

Adi Shankaracharya says this Universe is equal to a City seen in a mirror. What we see inside the mirror as City is not the reality. It is 'reflection' of the City. The City has buildings and lanes big and small. The city has people moving all around. All these are inside (our) mirror. But neither the sizes, shapes, characteristics of what see in the mirror are original characteristics of the City or its streets or its people.

What we understand from the image of the City in the mirror is our perception. What the real city that exists outside is, the reality. There is this duality here, which means perception and reality are two different things in this case. This duality, which is perception and reality being different, is mAyA or illusion or ignorance.

Similarly, what we see as Universe is not reality. What we see as Universe is perception. What exists is reality. There is a difference between our perception of the Universe and its reality. There is duality.

9 THE SLEEP OF MAYA

This illusion or ignorance of mAyA that causes the duality (of perception and reality being different) is one seeing their own (perspective), totally internally when they are sleeping.

When we are sleeping, we can perceive only what's inside of our mind. We do not see the outside reality. The outside reality arises only when we wake-up. What we see inside in our sleep is called a 'dream'. It is only a perception. It is not reality. There is duality here as perception (sleeping) and reality (waking up) are different.

So, our understanding of the Universe is like us seeing totally internally in a sleep with duality of perception and reality being different.

10 ATMAN, THE GURU, WAKES US UP

So, who can wake us up and make us understand the Universe and its reality...?

Atman can wake us up from this sleep. When does Atman manifest...? When our thoughts become detached observers or witnesses of our bodies and actions as much as others, then Atman is said to manifest in us.

With this detached observer in us, which is the Atman in us, we become awake from this sleep of mAyA. We get awakened from illusion or ignorance of self-perspective. With witnessing thoughts, we are able to analyze and understand multiple perspectives. We overcome the mAyA.

The detached observer in us that awaken us from our ignorance is our Guru because it is the duty of a Guru to awaken the knowledge in us. Since these witnessing thoughts are a manifestation of the Atman, Atman is the Guru that wakes us up. According to Sankara, this Atman is dakShinamUrthy from whom the Universe got projected, as a banyan tree is projected from a tiny seed.

11 DAKSINAMURTHY STOTRAM - 1

viśvaṃ darpaṇadṛśyamānanagarītulyaṃ nijāntargataṃ

paśyannātmani māyayā bahirivodbhūtaṃ yathā nidrayā

।

yaḥ sākṣātkurute prabodhasamaye

svātmānamevādvayaṃ

tasmai śrīgurumūrtaye nama idaṃ śrīdakṣiṇāmūrtaye ॥

1॥

Universe (visvam) is equal (tulyam) to a city (nagari) seen (drsyamana) in a mirror (darpana). Seeing (pasyan) totally internally (antargatam) one's own (nija) mAyA/illusion like sleep (yatha nidraaya), outside (bahri) is like (iva) arising (from sleep) (udbhutam) the time of awakening (prabodha samaya), our self (sva atmanam) does/performs (kurute) the actual (sAksat) with advayam (non-duality). Unto that Sri Guru mUrthy, salutations to that Sri dakShinamUrthy.

Observer is the Guru

Here are a few snippets from Guru Gita4, in which Shiva instructs Parvati on who is 'Guru'...?

12 DEFINITION OF A GURU

The conventional translations often say Guru is Shiva or Guru is brahman, Guru is the Self and serving Guru is the way to cross the ocean of manifestation. It kind of sends a message that any Guru needs to be served at feet, ablutions to guru be sipped and sprinkled on head etc. Essentially Guru Gita has been used to highlight how all one should revere a Guru. But in my understanding, Guru gita actually defines who a 'Guru' is. Not anyone can be a Guru. A Guru needs certain qualities.

Guru gita says Guru is like the devas completely devoted to the 'far-off' (parA). parA is Atman. For example, Shiva (maha-devA) is a Guru, as Shiva is devoted to Atman. Making Guru a mortal is like creating sexual connections between mortals and immortals. But the learning of Vedas, sAstras, purAnas, itihaAsas, mantra-tantras, shaiva shakta agamas confuses the minds of beings. If one does not understand the Guru-tatvvam all the Japa (meditation), Tapa (penance or hard work), Vratas (vows) and tirtha-yAtras (pilgrimages) become wasted. Guru is none other than buddhi-Atman that creates the efforts for our thoughts and actions. Buddhi Atman is the Atman that facilitates the evolution of buddhi which drives our thoughts and actions. Atman is the detached witnessing thought in us that remove the duality of perception and reality by making us move away from self-perspective (maya) and the resultant illusion (mAyA), thus facilitating knowledge evolution. The hidden avidhya in us causes the self-perspective induced illusion (mAyA) of attachment to our body (deha) and results in ignorance.

VijnAna is the clarity (prasAda) expressed by the sounds of the Guru (the detached witnessing thought).

Guru is the observer self in us that drives our knowledge evolution by removing the duality of perception and reality (dvaitam) and bringing in the one reality before us. Hence a two-legged person who can remove the perception (duality) in us (like the detached self in us), gives us knowledge to cross the ocean of manifestation is a 'guru', at whose feet we should serve, whose ablutions we should drink and sprinkle on our head. So it is not just any 'Guru'. The real Guru for all of us is our detached/observing self, which is the manifestation of that Atman in the bodies. An external person who can do the same as our observing self, can also be considered a Guru and only such a Guru who is like our observing self has to be deemed as 'Guru'.

13 GURU - THE MASSIVE ONE

Guru means the 'heavy' or 'massive' one. Any heavy or weighty objects are called Guru. For example, the planet Jupiter or Brihaspati is called Guru, because it is the largest planet in our Solar system with most mass.

When witnessing thought manifests in us, Atman is said to manifest in us and that Atman is called Guru, the massive one. The witnessing thought detached from the body and its actions actually relates to all the bodies and their actions in the Universe. Hence this detachment with our own becomes attachment with everyone. This attachment with entire Universe leads to jnAna or wisdom or knowledge. Hence this witnessing thought is said to be the 'Guru', the heavy one.

14 SUMMARY OF VERSES 17 TO 28

Mahadeva says to pArvati

17. The secret of extreme secrets is not fit to be spoken/said/revealed. It is not said/explained before anyone. Due to your bhakti, I explain it to you.

18. You are my form, therefore I am saying these. Why these questions assisting the world, not even asked before...?

19. Guru, similar to devas are devoted to the parA (Atman). These are already spoken and understood by the great minded people (with great thoughts)

20. Guru is said to be Shiva. Shiva is remembered as the Guru. Any difference becomes making Guru a mortal force with whom physical relationship is possible.

21. Guru is brahman which manifests as self (aham) and none other. (It is not a mortal force with which physical relationship is possible). This is scarce in the three lokas and I am telling you that.

22, 23. The Guru is brahman, which is our 'self' in reality. But the learning of vedas, sastras, purAnas, itihasas, mantra, yanta, occults, agamas of shaiva/shakta and other multiple paths and their mixing up, confuses the mind of all living beings.

24. If this Gutu tattvam of our Guru being the brahman manifesting as our self is not understood, all the penance, meditation, travel, sacrifices et al become wasted.

25. O lovely faced radiating woman, the knowledge of self is Guru, (which) becomes striving actions and thoughts and nothing else is truth. Our Self, the Guru, the controller of our knowledge, drives our actions and thoughts

26. The hidden unlearning causes deha-mAya and ajjAna. The words of Guru causes the VijnAna. How...?

'ma' is 'me' or 'my' or 'self'. 'maya' is mine or a self-centric perspective. mAya is the illusion that comes out of this self-centric perspective. This maya, the self-centric perspective arises from the attachment of thoughts to the body of the beings. This is the 'deha-mAya' of all beings (jagat).

What causes this attachment of thoughts to the body...? It happens when Vidhya is hidden. Vidhya is learning. Our thoughts manifest in the body and perform actions through the body. Hence, we learn to attach our thoughts to the body. We develop mAya the illusion due to this attachment with this body.

But we need to unlearn this attachment of thoughts to the body or this self-perspective. We need to develop observing/witnessing thoughts not attached to our body. Only when we unlearn this self-perspective, we can grow out of mAya. When this unlearning is hidden, we remain in mAya and attached to the body.

By the prasAd of words spoken by the guru, the self, vijnAna happens. Guru, the self teaches VijnAna. With VijnAna, the duality of perception and truth being different disappears. An observing self, detached from the body, brings us much closer to reality and removes the perception bias. Thus, non-duality of being closer to reality alone (rather than perception) appears.

27. Whose dual lotus foot restrains, burns duality, delivers/crosses the river/ocean of manifestation, to that guru,the self, bows down/salutes. Shiva is Guru. Brahman is Guru. The knowledge of Atman is Guru. That manifests as our 'self' without attachment to body is Guru. This Guru, the detached self helps overcome mAyA, teaches VijnAna. Thus, a two-legged person who restrains and burns the duality, delivers across the ocean of manifestation, like our 'self, is a Guru.

28. Out of kindness for you, let me explain that brahman exhibits/gives the manifestations of which those with self (Atman) pure of all pApa serves at the feet of such a Guru.

15 Sloka 17 to 28 of Guru Gita

shrI mahAdeva uvAcha |

na vaktavyam idaM devi rahasya atirahasyakam
na kasyApi purA proktaM tvad bhaktyarthaM vadAmi tat || 17||

Not fit to be spoken/said (na vaktavyam) here (idam) devi secret of the extreme secrets (rahasya ati rahasyakam). Not (na) of whom/whose (kasya) even (api) before (purA) is said (proktam) due to your bhakti (tvad bhaktyartham) I say/explain (vadami) that secret (tat)

mama rUpAsi devi tvam atastat kathayAmi te
loka upakArakaH prashno na kena api kRtaH purA || 18||

You (tvam) are of my (mama) form (rupasi) devi, therefore (atastat) I am saying these (kathayAmi te). Questions (prashno) assisting/helping (upakAraka) people of the world (loka), na (not) done (kRtaH) before (purA), even (api) why (kena)

yasya deve parA bhaktir yathA deve tathA gurau
tasyaite kathitA hy arthAH prakAshante mahAtmanaH || 19||

On whom/which/whose (yasya) the far-off, often used to denote Atman (parA) devas which are immortal divine forces (deve) are devoted (bhaktir), like those devas (yatha deve), same way the gurus (tatha gurau). All these (tasyaite) are already spoken (kathita), its meaning (arthah) revealed (prakashante) to the great minded or those with great thoughts (mahAtmanah)

yo guruH sa shivaH prokto yaH shivaH sa guruH smRtaH
vikalpaM yastu kurvIta sa naro guru talpagaH || 20||

This guru (yo guru) is said to be shiva (prokto shivah), that shiva (yah shiva) is remembered as the guru (sa guru smRtah). Any difference (vikalpam yastu), becomes (yastu) making (kurvita) guru (guru) mortal (nara) with sexual intercourse (sa talpaga).

durlabham trisu lokesu tac zrnusva vadamy
aham guru brahma vina na anyau satyam satyam
varAnane || 21||

That which is scarce (durlabham) in the three lokas (trisu lokesu), hear me (zrnusva)I tell you that (tac vadamy), self (aham) guru (aham) is none other than (vina na anyau) brahman, truth, truth oh lovely faced (satyam sayam varanane).

veda sAstra purAnAni ca ithAsAdikAni ca mantra
yantrAdi vidyAnAm
mohano ucchAdanAdikamshaiva shAktu Agamadini hy
anye ca bahavo
matAuapabrahmsau samasthAnAm jivAnam bhrAmta
cetasAm|| 22|||| 23||

The learning of Vedas, sAstras, purAnas, itihAsas, mantra, yantra, occults/tantric (mohana ucchadana), shaiva agama, shakta agma and other multiple (anya bahava) religions/paths (matau) and mixing up of all these, confuses (bhrAmta) the mind/behavior of jivas/beings (samasthanam jivanam cetasam).

japas tapo vratam tIrtam yajno dAnam tatha iva
guru tattvam avijnAya sarvam vyartam bhavet priye||
24||

Japa, tapa, vratam, tirtha yatras (tirtam), yajna, dAna and all of these (sarvam) become wasted (vyartam), if one is not aware of knowledge (avinAya) of guru tattvam.

guru budhhyAtmano na anyat satyam satyam varAnane
tallabhArtam prayatna astu kartavya ca manishibhih ||
25||

Guru is none other (na anyat) than the concept/knowledge of self (buddhi atman) , truth, truth, o lovely faced, radiating woman, becomes striving (prayatna astu) actions and thoughts/intelligence.

gUdha avidhya jagan mAya deha ca ajjAna sambhava
vijjAnam yat prasAdena guru shabdena kathayate || 26||

The hidden avidhya (gUdha avidhya) causes (sambhava) the deha-mAya of beings and ajjAna. VijnAna, which is by prasAd (yat prasAdena) of the guru/self's words (shabdena), that are spoken (kathAyate).

yada~NghrikamaladvandvaM dvandvatApanivArakam
tArakaM bhavasindhoshcha taM guruM
praNamAmyaham || 27||

Which/what (yad) lotus foot/root (anghri kamalam) dual (dvandvam) restrains (vArakam) burns (tApani) duality (dvandvat), delivers/crosses (tArakam) ocean/river of manifestation (bhava sindhu) that guru (taM guru) self (aham) bows down (pranamam)

dehI brahma bhaved yasmAt tvat kRpArthaM vadAmi tat

Sarva pApa vishuddhAtmA shrIguroH pAdasevanAt || 28||

Out of kindness for you (tvat krpa artham) explain that (vadami tat), from which/whom (yasmAt), brahman exhibits/gives (dehi) the manifestation (bhaved) from which (yasmAt) Atman cleaned/pure of all pApa (sarva pApa vishuddha Atma) than the service at the feet (pada sevanat) of the Guru (shri guru)

Observer drives evolution

Sa-akSi means 'that eye'. It is the 'witnessing' eye, the eye of 'third party' that is witnessing, but not interacting with what's happening or not changing due to the interaction. It is that which facilitates the interaction but, in the end, do not become part of the interactions and stands alone as an observer or witness to that interaction. Such an observer is a 'Guru'.

This observer or witness drives the evolution of the Universe. In every piece of Universal evolution, there is a third-party observer, who at the end stands out of the interactions after facilitating it. Without this observer, evolution will not happen.

Our Universe can be classified into four domains from the perspective of how the laws operate. They are quantum domain in which force-fields and particles evolve, classical domain in which matter evolves, biological domain in which life evolves and consciousness domain in which thoughts or information exchange in living beings evolve. Observer driving the evolution is a common law across all the domains.

16 Dark Matter as observer

Cold Dark Matter (CDM) dominates the Universe when matter evolution started. This Cold dark matter is the 'observer' of the Interacting matter as defined above. Interacting matter acquires and releases energy through electromagnetic interaction that causes creation and destruction of large structures of matter in the form of stars, planets, and biological life. The Cold Dark Matter does not acquire or release energy through electromagnetic interaction and hence does not build or evolve like matter and also does not interact with matter. It is estimated to be 3 times massive as interacting matter.

Since Cold Dark Matter is massive it bends spacetime causing huge gravitational impact on matter, binds matter together and makes matter evolve. Though Cold Dark Matter itself may be changing in shape as the matter evolves, but does not become part of the evolution of matter. Cold Dark Matter stands out as it is, just as an observer or witness as matter evolves into its more complex forms. This evolution of the Interacting matter is described as the 'sacrifice' or 'Yajna' on the 'bed' of Cold Dark Matter.

Though the Cold Dark Matter just remains as an observer/witness/sAkshi, without it matter cannot evolve in the way it has evolved, though matter can exchange energy through Electromagnetic interaction. Since Cold Dark Matter and Matter are both 'massive' matters, CDM is like our manas-sAkshi, the observer in us that evolves our thoughts.

17 NEUTRON AS OBSERVER

Atomic nucleus has neutrons and protons. Protons interact with electrons and build up larger elements, compounds, long polymer chains and the biological life even. But neutrons remain just observers or witness in the nucleus. They do not interact with electrons or protons.

Though neutron is only an observer/witness/sAkshi, it facilitates the binding of positive charged protons into each other. Without the neutrons, the protons would repel away from each other. But neutrons reduce the electromagnetic repulsion, bind the protons and make it into an atomic nucleus, thus bringing the next stage of evolution. Neutron and Proton are part of the same atomic nucleus. Hence this is also similar to our manas-sAksi, the observer in us that evolves our thoughts.

18 CATALYST AS OBSERVER

In Chemical or biological processes, be it synthesis of proteins or DNA replication, catalysts are involved. Catalysts are those that facilitate the evolution, that do not become a part of the evolution, stands apart from the evolution, after facilitating the evolution.

For eg. enzymes facilitate the evolution of substrates by binding to them, undergoing conformational or shape changes to induce fit with the substrates. As substrates evolve into different forms, enzymes stand apart just as a witness or observer to that evolution. Catalysts are part of the same system whose evolution they drive or propel. Hence catalysts are said to be similar to our manas-sAksi, the observer in us that evolves our thoughts.

19 DARK ENERGY AS OBSERVER

The Universe's spacetime is continuously expanding. The energy behind this expansion is called 'Dark Energy'. Both dark energy and energy are 'energy' characterized by momentum and doing some 'work'. The dark energy does not contribute to building up the matter. It is contributing to the expansion of the spacetime fabric of the Universe.

Hence dark energy is only a witness or observer to the energy and its evolution into different forms in the Universe. Hence dark energy as observer of energy is similar to our manas-sAksi the observer in us that evolves our thoughts.

20 OBSERVER DRIVES OUR EVOLUTION

When human beings observe the 'self', become an observer/witness for themselves, their bonding become detached, balanced and happy. They become catalysts for driving their own evolution into the next stage of evolution.

One can truly be the witness or 'unchanging observer' of one's own, but not others. So, one has to develop being a total independent observer of oneself. This is what ashtavakra gita also says. Be an observer of yourself. The detached, unchanging 'observer' drives our evolution by overcoming mAyA. Hence it is the 'Atman' or 'I' or 'self'.

21 DESIGN PATTERN OF EVOLUTION

There seems to be a design pattern in the evolutionary processes across these domains.

1. There is puruSa which facilitates evolution, though it does not contribute to the energy for an evolutionary process.

2. There is prakRti, the object of evolution, which keeps evolving. All the energy for its evolutionary process comes from within prakRti itself.

3. puruSa facilitates the evolution of prakRti by binding different parts of prakRti, bringing it together and enabling different parts of prakRti to interact/react with each other and evolve.

4. In this process, puruSa undergoes shape changes (or confirmatory changes) to induce a fit so that it binds with prakRti and brings different parts of prakRti together to evolve.

5. As the evolution of prakRti proceeds, puruSa remains as it was before without any evolution, while prakRti has changed forms, shapes and sizes. Because puruSa stands apart from this evolution it is said to remain as an observer or witness of the evolution of prakRti.

6. prakRti is said to be 'blind' to puruSa's role in its evolution, as all the energy requirements for the evolution of prakRti comes from prakRti itself.

7. puruSa is said to be 'crippled' or 'halted' (equated to being 'lame') because it does not absorb or emit energy in an evolutionary process and remains as it is.

8. saMkhya kArika says in the union of blind and crippled the creation flows.

The design patterns

The observer or witness facilitating the evolution of something, but standing apart from it as an observer or witness is the principal component of saMkhya in saMkhya kArika[5]. saMkhya is understood as a philosophy. It is not. It is a 'design pattern' for Universal evolution.

saM+khya means making known everything or knowledge of everything. What 'makes known everything' is a 'design pattern'. A design pattern is a repeatable or re-usable form of solution or a pattern that can be used across multiple domains.

saMkhyA is the design pattern as seen in the evolution of the Universe. saMkhya builds up on this observer/witness driving evolution and propounds a design pattern in more detail.

saMkhyA calls the facilitator of the evolution which does not become part of the evolution, but stands apart as an observer or witness of the evolution as 'Purusha'. puruSa means that is 'before (pur) the dawn (uSa)'. Since the facilitator of evolution exists before the 'dawn' of any evolution, it is called puruSa.

22 THE 24 PRINCIPLES OF SAMKHYA

saMkhya kArika explains the principles of saMkhya.

saMkhyam prakurvate ca iva prakRtiksa pracakSate
Catur vimsati tattvani tena sAMkhyAh prakirtitah

The doer/maker (prakurvate) and 'that is being done/made' (prakrti) it is said (pracaksate) are of 24 principles (catur vimsati tattvani) that (tena) is the declaration of sAnkhya (sAnkhyah prakirtitah).

Anything that is created/made with a creator/maker involves 24 principles says saMkhya. This is the design pattern saMKhya sees across all creations.

23 THE ROOT-PRAKRTI IN EQUILIBRIUM

Mūlaprakṛtir avikṛtir, mahadādyāḥ prakṛtivikṛtayah sapta |
shodashakas tu vikāro, na prakṛtir na vikṛtih puruṣah |3.|

The 'starting/root prakrti' (mUla-prakrti) does not evolve. prakRti means that first/foremost (pra) of creation (kRti). This creation is in equilibrium or balance. Hence it does not change. Hence this foremost creation, prakRti is called avikRti. vikRti means that changes or evolves because it is not in equilibrium or in imbalance. avikRti is that does not change or evolve because it is in equilibrium or balance.

vikRti means the evolved or differentiated form. vikAras are the modifiers. From the balanced prakRti, seven differentiations (sapta vikrti) happens and sixteen modifiers (shodasa vikara) emerge. Purusha is not prakrti (in balance) or vikrtri (imbalanced). It just remains as an observer or witness.

Thus, the seven differentiations, sixteen modifiers and one prakrti becomes the 24 principles of saMKhya, the design pattern of Universe as well as human beings. Of the sixteen modifiers, one is Purusha, who is just a witness or observer or catalyst that stands alone from prakrti.

24 PURUSA FACILITATES EVOLUTION

In comparison, the puruSa is neither prakrti, the foremost creation in equilibrium nor vikrti, the one that does change or evolve because of non-equilibrium or imbalance. puruSa remains an observer or witness to the evolution of prakRti.

PuruSa facilitates the evolution of prakRti by binding different parts of prakRti, bringing it together and enabling different parts of prakRti to interact/react with each other, thus changing its equilibrium. This results in the evolution of prakRti. But puruSa does not become part of prakRti. It stands apart from prakRti just as a witness or observer to prakRti. This 'pattern'. happens in many domains.

This is explained in saMkya kArika sloka 21

Puruṣasya darshanārthaḥ, kaivalyārthas tathā pradhānasya |
pangvandhavad ubhayor api, saṃyogas tatkṛtaḥ sargaḥ |21|

Exhibition (darzana) of puruSa to arthah, while standing apart from artha, similarly of pradhAna, from the union of this crippled and blind, creation happens. prakRti is the subject of evolution under equilibrium. Artha is the subject of evolution as it is evolving.

puruSa is a facilitator of evolution of artha. It brings different parts of artha together and binds them. When it does so it becomes the 'Principal' mover of evolution and hence called pradhAna. Both puruSa and it's another form pradhAna stand apart from artha.

Artha is blind to puruSa as it does not use the energy of puruSa for its evolution. puruSa is crippled as it cannot give out or acquire energy like artha and hence remains as it is during the evolutionary process. From their union, creation happens.

25 PuruSa as observer

The second sloka of Vishnu SashasranAma says

pūtātmā paramātmā ca muktānāṁ paramā gatiḥ |

avyayaḥ puruṣaḥ sākṣī kṣetrajñōkṣara eva ca || 2 ||

The pure (pUta) and omnipresent (parama) Atma and the final destination (parama gatih) of liberation (muktAnam) is the non-decaying (avyaya) witness/observer/the-eye (sAksi) the knower of the field/place (kSetra jna) and indivisible (akSara)

Increasing entropy/no. of microstates	Increasing transactional energy	Increasing mass
Information content or Knowledge	Power to exist independently and move	Grow to larger structures, reproduce
Sattva	Rajas	Tamas

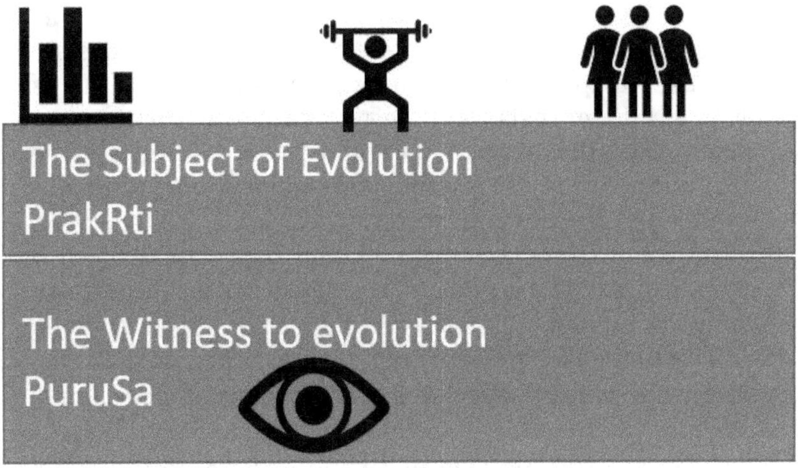

puruSa is the undecaying, indivisible knower of the 'field' on which matter and beings evolve. This puruSa is manifestation of that Atma which is pure, omnipresent and final destination of all liberations. kSetra is a field or place on which growth or change happens. It is equivalent to prakRti. This is the purport of the given sloka. The above figure illustrates it.

26 PURUSA CAUSES ENTROPY CHANGE

The starting point of this creation is 'mahat'. Mahat is the differentiation in prakRti caused by an increase in entropy due to the 'witnessing' of puruSa.

Entropy can be understood roughly as equivalent to the number of possible microstates or 'information content' of a system. When matter structures become larger or more complex, entropy is said to increase.

At the start of Universe, there was Quark-Gluon-Plasma (QGP) in which quarks and gluons exist in a 'de-confined' state at very high temperatures. It cannot evolve further. The QGP is in equilibrium or balance if left alone. The dark energy expands the Universe's spacetime, which reduces the temperature of the QGP system. This results in QGP system to transition into a system of hadrons. This is called confinement. This transition from a system of quark-gluon-plasma to a system of multiple hadrons is equivalent to an increase in entropy or number of microstates or information content. Thus, due to the facilitation by dark energy, the entropy in the system is increased. Dark Energy that expands the spacetime stands apart as an observer to this evolution, as it stands apart from the energy in the QGP or Hadron system. Dark Energy also does not contribute to this transition in terms of energy. Hence energy is blind to Dark energy. Dark energy is crippled as it does not evolve like energy during the evolutionary process.

The Cold-dark matter bends the spacetime more than what matter itself does (as cold dark matter is more abundant than matter), brings matter together, make matter interacts and evolves it further. Cold Dark Matter stands apart as an observer or witness to the evolution of matter. It also does not contribute any energy to matter evolution. Hence matter is blind to Dark matter. Dark Matter is crippled as it does not evolve like matter during the evolutionary process.

Catalysts often bind with chemicals providing them a platform to interact, facilitating their interaction and new chemicals forming, while catalysts themselves stand apart as observers or witness to the evolution of Chemicals. They do not contribute any energy to the evolution. Thus, the chemicals are blind to catalysts, while catalysts are crippled as they do not evolve during the evolutionary process.

In biological terms, entropy or number of microstates or the information content, becomes knowledge. Viruses have genetic material that can be replicated, but lack life-sustaining processes that can replicate them. Thus, viruses have knowledge to be transmitted in the form of genetic material as the genetic material has the knowledge to encode proteins. This creation of genetic material is an increase in entropy, the mahat, compared to just organic compounds. Hence Viruses have 'mahat' an increase in entropy. But Viruses cannot transfer knowledge themselves to others as they lack life-sustaining processes. Hence, they lack ahamkAra, the internal property that allows them to build and evolve further on their own.

In all these cases, some 'X' upsets the equilibrium of 'Y' and increases its 'entropy' or number of microstates or information content, though X does not contribute energy to Y. X also stands apart from Y as Y keeps evolving.

Be it Hadrons, new matter forms or new chemicals or viruses the entropy or information content of the resultant system is more evolved in terms of information or knowledge than the previous one. This increased entropy is called the 'mahat'.

27 THE SEVEN DIFFERENTIATIONS

Starting from mahat, the sloka 3 of saMkhya kArika says the creation undergoes seven imbalance or non-equilibrium or differentiations (mahadādyāḥ prakṛtivikṛtayah sapta).

Samkhya kArika sloka 22 says

Prakṛter mahāms, tato ahamkāras, tasmād gaṇash ca shodashakaḥ |

tasmād api shodashakāt, pañcabhyaḥ pañca bhūtāni

From prakRti, mahat, then ahamkAra and then set of sixteen. Then from the set of sixteen from the sets of five, the pancha-bhutas.

These are the seven differentiations

- Mahat is differentiation caused by an increase in entropy as seen above.
- aham-kAra is the self-property or self-sustaining property that allows to build and evolve further.
- The five pancha-bhUtas are the differentiation caused by different states of matter.

28 MAHAT TO AHAM-KARA

In the Universe's terms matter forms arising from an undifferentiated Quark Gluon Plasma into differentiated 'heavy hadrons', with increasing entropy is mahat. But they lack self-sustainability and hence lack 'aham-kAra'.

In biological terms, the organic compounds which hitherto did not have 'signaling' or exchanging of information, developing signaling system or exchanging of information, with increasing entropy is mahat. For example, viruses have signaling systems, but do not have self-sustaining life processes. Hence, they lack a 'aham-kAra'.

In Universe's terms, such matter forms like 'heavy hadrons' that arise with increased entropy do not live long. They are subjected to weak force and decay. Thus, they do not have the property of self that allows them to build and evolve further. The first matter forms that arise with property of self that allows them to build and evolve further are the neutral atoms. Atoms are said to have ahamkAra as they exist independently, transact energy and evolve further. This aham-kAra of atoms causes all the matter forms and subsequently evolution to biological beings.

In biological terms, primitive organisms that can sustain themselves with life sustaining processes such as bacteria have aham-kAra, the property of self that allows them to build and evolve further into more and more complex beings.

29 AHAM-KARA TO BHUTA AND PRANA

From this property of self on which creations build and evolve (the ahamkAra), 'massive' matter gets differentiated into five principal states. They are the pancha-bhUtas in Universe or pancha-prAnas in biological beings.

In the Universe's terms, matter gets principally differentiated into Solids, Liquids, Gas, Plasma and Bose-Einstein Condensate. These are the five principal states of matter or the pancha-bhUtas.

- Solid makes the 'Earth' and is called 'prithvi'.
- Liquid makes the 'water' and is called 'Apa or 'Jal'.
- Gas makes the 'air' (or more precisely mixture of compounds) and is called 'vAyu'.
- Plasma makes the 'fire' and is called Agni (through which energy is transferred from one to another).
- Bose-Einstein Condensate makes the matter in interstellar space of the Universe and is called 'AkAsa'.

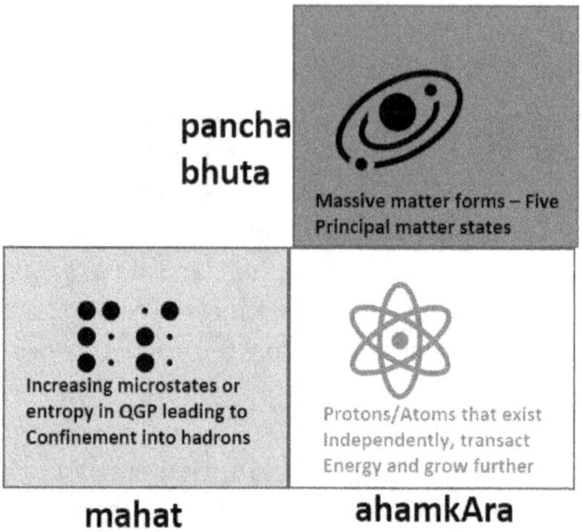

The union of these different states of matter creates all the matter structures and even the biological beings of the Universe

In biological terms, biological functions get differentiated into prAna, samAna, apAna, vyAna and udAna. These are the five principal functions of biological beings or the pancha-prAnas.

- prAna is the respiratory system
- samAna is the digestive system
- apAna is the excretory system
- vyAna is the circulatory system
- udAna is the nervous/signaling system

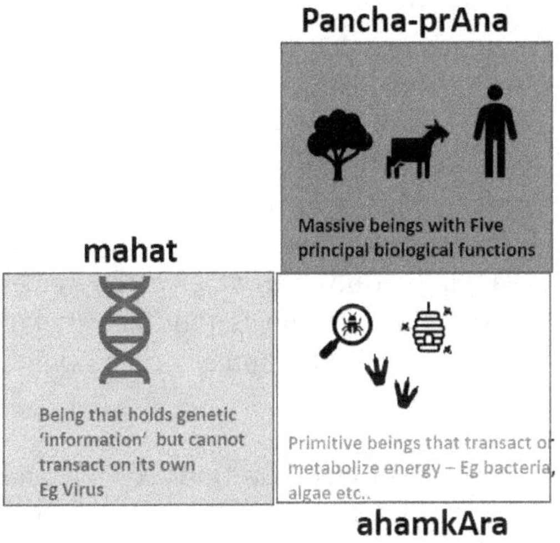

If we visualize the Universe as a human being,

- The movement of molecular clouds or vAyu (air) across the space which indicates Universe is breathing/active is the prAna or 'respiratory system' of the universe.
- Stellar nucleosynthesis processes like alpha, triple-alpha, Carbon-Nitrogen-Oxygen (CNO) which create newer elements are the digestive fire/system or samAna of the Universe.
- The 'solid' matter (prithvi or earth) thrown out from these processes which becomes the base for further evolution (of say biological beings) is the apAna of the Universe.
- The movement of energy in the form of light or radiation across the space of the Universe is its circulatory system or blood. It is the 'celestial water' or 'Apa' (Jal) of the Universe.
- The matter in the interstellar space or AkAsa is in the Bose-Einstein Condensate state. It is the udAna, the signaling system of Universe on which other beings evolve.

If we visualize human beings as the Universe,

- prAna the respiratory system is the movement of molecular clouds which is the vAyu.
- samAna, the digestive fire are the stellar nucleosynthesis processes, which is the agni.
- apAna, the excretory system are the excreted solid matter forms that make the prithvi, the earth.
- vyAna, the circulatory system (blood), is the radiation of light through the Universe, which is the Apa or Jalam.
- udAna the signaling system is the 'interstellar space' that connects all the organs and all organs evolve out of that system or space.

Thus pancha-bhUtas form the Universe. Pancha-prAna form the biological beings. Since both have the same design pattern, human beings are said to mimic Universe and Universe is said to mimic the human beings.

The mappings given above for the pancha bhutAs and pancha-prAnas are traditional but not understood uniformly in the same way by all. Here I have used one particular type of mapping.

30 THE SIXTEEN MODIFIERS/VIKARAS

Samkhya kArika sloka 24 says

Abhimāno ahaṃkāras, tasmād dvividhaḥ pravartate

sargaḥ

ekādashakash ca gaṇas, tanmātra pañcakañ caiva |24|

From ahamkAra, the creation progresses in two ways. A set of eleven and the five tanmAtras. It further says in Sloka 26 and 27

Buddhīndriyāṇi cakṣuḥ,

śrotraghrāṇarasanatvagākhyāni

vākpāṇipādapāyū, upasthāḥ karmendriyāny āhuḥ |26|

The indriyani of Buddhi are known as (Akhyani) caksuh (sight), zrotram (hearing), ghrAna (smell), rasana (taste), tvac (skin/touch). The Indriyani of Karma are vac (speech/expression), pAni (working by hands), pAda (movement by foot), pAyu (metabolism involving excretion) and upasthah (procreation or reproduction). These are the ten indriyas.

Ubhayātmakam atra manaḥ saṃkalpakam indriyaṃ ca sādharmyāt
guṇapariṇāmavisheshān, nānātvam bāhyabhedāsh ca |27|

Both Atma, in this place according to the determination of mind and indriyas, from the dharma (the laws of the Universe), causes specific guna evolution and multifarious external differentiations.

The Purusha, the facilitator of evolution, which ultimately is not part of the evolution, but stands apart as observer of the evolution is the Atman. Atman is the detached, witnessing/observing thoughts, witnessing itself and others and that observer causes the changes in gunas and all the external differentiations.

The thoughts attached to the body are the 'manas'. The thoughts detached from the body is the manifestation of Atman.

Thus, the ten Indriyas and the Purusha are the eleven modifiers apart from the tanmAtras. Thus the sixteen modifiers are

- pancha buddhi indriyas that modify mahat, the entropy or buddhi
- pancha karma indriyas that modify ahamkAra

- pancha tan-mAtras that modify pancha-bhUta or pancha-prAna
- puruSa, the observer/witness that modifies all

The above sixteen are the sixteen modifiers following the same design pattern in Universe's terms or biological terms.

31 Five modifiers of Entropy

In Universe's terms, the entropy (or microstates or information content) of the Universe is impacted by five forms of energy. These energy forms add or reduce the microstates or information content of any system. They are the buddhi indriyas. They are

- Electromagnetic energy
- Mechanical energy
- Thermal energy
- Chemical energy
- Quantized states of atomic/molecular energies.

In biological terms, the buddhi/knowledge of a person is impacted by the senses of

- Sight
- Hearing
- Touch
- Taste
- Smell

The five energies and five senses are equivalent. Both impact the entropy.

- Electromagnetic energy causes sight.
- Mechanical energy causes hearing.
- Thermal energy causes Touch.
- Chemical energy causes taste.
- Quantized energy states cause smell.

Thus, the pancha buddhi indriyas of Universe and biological beings follow the same design pattern.

The mappings given above for the buddhi indriyas in biological terms is traditional and well understood. But the mappings given in terms of Universe's terms are my attempts at the mapping.

32 Five modifiers of energy

In Universe's matter terms, the ahamkAra (internal property that allows matter or beings to build and evolve further) of the Universe's matter is impacted by five functions which build and evolve the matter or being. These functions bind and build more and complex forms of matter. They are the karma indriyas. They are

- QCD binding leading to hadrons
- Nuclear binding leading to nucleus
- Atomic binding leading to atoms
- Gravitational binding leading to gravitationally bound matter forms
- Molecular binding leading to molecules

In biological terms, the ahamkAra of the beings is impacted by these five functions. These functions modify build and evolve the biological beings.

- upasthA (procreation)
- pAyu (metabolism involving excretion)
- pAni (working by hands)
- Vac (Speech)
- pAda (physical movement on foot)

The five binding functions and work organs are equivalent. Both impact growth.

- Upastha (procreation) is the QCD binding that creates the first hadrons.

- pAyu, the metabolism involving excretion is the nuclear binding that grows the atomic nucleus larger and larger (builds mass).
- pAni the working by hand is equivalent to atomic binding that builds electron orbitals around an atomic nucleus creating atoms.
- Vac, the human speech is equivalent to expression of different forms of matter through molecular bonds.
- pAda the locomotion by foot is equivalent to gravitational binding, as gravitational binding moves matter across spacetime.

Thus, the pancha karma indriyas of Universe and biological beings follow the same design pattern. The mappings given above for the karma indriyas in biological terms is traditional and well understood. But the mappings given in terms of Universe's terms are my attempts at the mapping.

33 THE FIVE MEASURES

tanmAtras are 'measures' that modify the pancha-bhUtas or pancha-prAna, like buddhi-indriyas modify mahat and karma-indriyas modify ahamkAra.

The sloka 38 of saMkhya kArika says

tanmātrāṇi avisheṣās, tebhyo bhūtāni pañca

pañcabhyaḥ

tanmātras are 'avisesa' which means 'not differentiating' or 'common properties'. The tanmAtras are common properties of the bhutA. They differentiate the bhutA into the pancha-bhutA or five states of matter.

In Universe's terms, the measures that are common to the pancha-bhUta are

- Density (Ratio of mass to volume)
- Velocity
- Pressure,
- Volume
- Temperature.

In biological terms, the measures that differentiate and modify the pancha-prAna are

- Oxygen saturation
- Heart-rate
- Blood-pressure
- Respiratory rate
- Body temperature

The five measures of pancha-bhUta and pancha-prAna are equivalent.

- Velocity is equivalent to heart-rate
- Pressure is equivalent to blood pressure
- Temperature is equivalent to body temperature
- Volume is equivalent to respiratory rate (volume of lungs)
- Density is equivalent to oxygen saturation (SPO2)

Thus the 'pancha' tanmAtras of Universe and biological beings follow the same design pattern. The tanmAtras are variously interpreted and no common understanding of them exists. The mappings given above for the tanmAtras in Universe or biological terms are my attempts at the mapping.

34 THE THREE TYPES OF MODIFICATION

While sixteen including the puruSa are modifiers, there are only three names/forms of modification from matter forms to our thoughts. The tri-gunas are the three names and forms of the evolutionary modification.

For example, all such evolutionary modifications in the Universal matter forms result in either change in entropy, energy or mass.

- When entropy increases even with the same energy, matter gets differentiated into different forms and is called 'mahat'. This increasing entropy is called 'Sattva Guna'.
- When the internal energy of matter forms a new entity that can not only exist independently but also grow and evolve further, it is said to acquire ahamkAra. This energy required for newer forms of matter to form is called 'rajas guna'.
- When the matter forms change in their mass, volume, pressure, temperature and velocity of motion, they become different states and become the pancha-bhutAs. The changing mass or becoming heavier or lighter is called 'tamo guna'.

Increasing entropy/no. of microstates	Increasing transactional energy	Increasing mass
Information content or Knowledge	Power to exist independently and move	Grow to larger structures, reproduce
Sattva	Rajas	Tamas

Thus, in Universe's terms, triguna are Entropy (sattva), Energy (rajas) and Mass (tama). In thoughts terms, the triguna are Knowledge/Information (sattva), Passion/Energy (rajas) and ignorance (tamas). The impact of the tri-gunas and their modification is seen across all domains of matter and beings, like wheels within wheels, which is explained with several examples in Chandogya Upanishad Chapter 6 later.

Samkhya kArika sloka 13 says

Sattvaṃ laghu prakāsham, ishṭam upashṭambhakaṃ
calaṃ ca rajaḥ
guru varaṅakam eva tamaḥ, pradīptavac cārthato vṛttiḥ
|13|

Sattva manifests/makes entities visible (prakAsam) easily (laghu). Sattva is the differentiation due to entropy or number of micro-states or information content or knowledge. The increasing entropy or information content that manifests or differentiates matter also causes interaction amongst the matter forms.

Rajas is the energy such that the entity is able to exist independently with specific properties and build/evolve more upon it. It also becomes the ability to 'exist and move' as an independent entity. Rajas is the property said to be desiring staying rooted (upasthambakam) or flowing (calam). Energy is either staying rooted as potential energy or flowing as Kinetic energy.

Tamas is the differentiation due to mass. It becomes the ability to become heavier (guru), larger and in turn be enveloping or enclosing (varaNakam).

The sloka says, these three gunas lighten up the expressions of 'artha' or matter forms. This means all the matter forms including the biological beings have their manifestation basis on the three gunas of Sattva, Rajas and Tamas. Thus the 24 principles of saMkhya and the tri-gunas are a design pattern across Universal matter forms, biological beings and even thoughts of human beings.

Tat Tvam Asi

35 ATMAN MANIFESTS AS PURUSA

Samkhya kArika says Atman, the observer or witness of evolution in Universe manifests as 'puruSa'. The dark energy, dark matter, the catalysts, the detached witnessing observer in our thoughts et al that remain a witness to the evolution of some other component are all different puruSas which are the manifestations of 'Atman'.

Manas are thoughts attached to our body and its actions. According to ashtavakra gita, Atman manifests as detached witnessing thoughts not attached to our body, but just a witness or observer like puruSa. Hence Atman is manas-sAkshi (witness of manas). To develop this observing component in our personality, the mAya or illusion caused by maya, the self-perspective has to be overcome.

The above design pattern is found across various ancient scriptures. Chandogya Upanishad explains how all the Univers's matter and beings evolve out of the 'tri-guna'. prakRti or the subject of evolution has three gunas or characteristics, which Chandogya Upanishad describes as 'Tejas', 'Apa' and 'Annam', which are same as 'Sattva', 'Rajas' and Tamas'. The evolution of prakRti kickstarts with Sattva or Teja becoming predominant. Then Rajas and Tamas become predominant one after another. Then again Sattva becomes predominant starting a new cycle of evolution in prakRti.

Uddalaka explains Svetaketu in Chandogya Upanishad, that Atman, the observer becomes the 'Satyam', the property of existence in all matter and beings as it drives the change of gunas and thus keeps evolving all matter and beings including consciousness and actions of beings.

36 CHANDOGYA UPANISHAD- 6

Chandogya Upanishad[6] Chapter 6 has the famous mahAvAkya 'tat tvam asi'. Chapter six explains events around svetaketu's learning.

Once upon a time there was Svetaketu, who was the grandson of Aruna. His father was sage Uddalaka. At twelve years of age of Svetaketu, his father Uddalaka sends him to learn from Gurukul saying all his family members had been educated in this way. Svetaketu spends 12 years in learning in Gurukul, till his 24 years of age and returns home.

Uddalaka sees a boy who is proud of his learning. So he asks him a question "Did your teachers teach you "through which unheard becomes heard, unthought becomes thought, unknown becomes known". Svetaketu is perplexed and asks his father for that teaching.

37 TRI-GUNA DESIGN PATTERN

Uddalaka gives an example. All things made of clay, gold or iron change form, but remain as they are. Likewise, existence changes forms, but at the core it is existence. All those existing came from a single existence.

Uddalaka says

vikaarah naamadheyam treeni roopaani iti; eva satyam

The truth is there are only three names and forms of modification. This assertion is same as in saMkhya kArika, where all modifications are said to be of three names and forms, viz, Sattva, Rajas and Tamas.

All modifications of existence have only three names or forms. They can be mapped to Tejah, Apah or Annam.

- Tejas means 'brilliance'. It is what makes something visible. It can be translated as fire or even increasing entropy or information content that makes new forms visible or manifest. This is the Sattva of saMkhya kArika.

- Apa is translated as 'Celestial Waters'. The 'Celestial waters' that sustain the evolution of Universe is 'light' or 'radiated energy', which when transacted between matter or biological forms, evolves them. This is the Rajas of the saMkhya kArika.
- Annam are the 'food grains' that convert energy to mass so that large populations can sustain on them. Energy to mass conversion sustains and evolves Universal matter forms. Annam, the food grains produces more food grains, like massive matter that produces more massive matter, converting energy into mass. This is the Tamas of the saMkhya kArika.

The words Teja, Apa and Annam are chosen from the view-point of human beings. What are the important needs for human beings? They are Fire, Water and Grains. The advent and controlled usage of fire was the initiator of human civilization. The knowledge of usage of fire differentiated human beings and laid the foundation of human beings of the future. Hence fire is equivalent to the Entropy or Information content or Sattva or Tejas for human civilization.

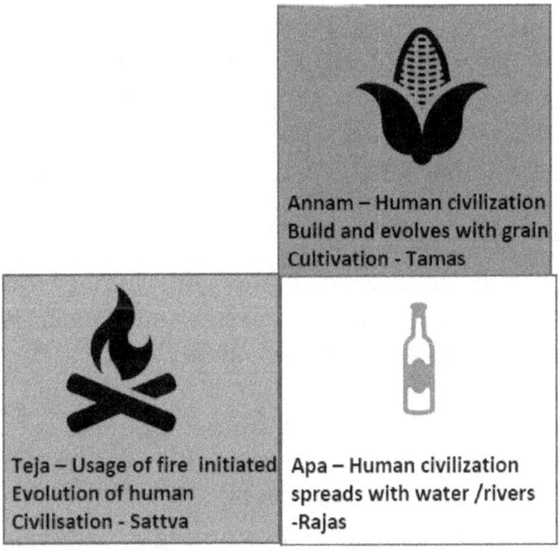

Human Civilizational needs

The next need for growth of human beings and their civilization was availability of Water. Human civilization thrived along the river valleys. This water is like Apa, the energy that spreads and helps sustain the Universe. If advent of fire created human societies, usage of water took them to next levels of growth.

The next need for growth of human being and their civilization was grain cultivation. This is the Annam, the food grains that have the carbohydrates and gives 'mass' to human population. The cultivation of food grains enabled human beings to settle down in a place and multiply rapidly to such an extent they conquered the entire planet. Thus, the civilization of human beings also follows the tri-guna design pattern.

Tejas or Sattva, which is increase in entropy or number of microstates or brillance or information content arose because of puruSa's observation which the prakRti cannot see as there is no energy contribution from puruSa. Hence Tejas is seen as arising on its own. Uddalaka gives an example of a person who has intensely desiring thoughts, leading to his perspiration. It looks as if the perspiration arose on its own but it arose from the person's thoughts. Teja (entropy) brings in Apa (energy or radiation). It is similar to the fact that when there is sun-shine, there is rain. Apa brings in annam (masses) everywhere. It is similar to the fact that when it rains, food grains are produced.

Uddalaka also gives examples of three forms of modification in terms of colors, Red, White and Black. Red is Tejas, the Sattva or entropy which is subtle. White is Apa, the Rajas or energy which spreads. Black is Annam, the Tamas, which is becoming 'more massive'. Another example is that of Adityam (sun) which is equivalent to red, Candram (Moon) which is equivalent to white and Krishnam (Earth) which is equivalent to black. The evolution of Adityam, Candram and Krishnam happens in the following way:

- The Universe evolved from QGP to atoms to massive molecular clouds, as described in the last chapter. The molecular clouds were Tamasic or massive and hence not evolving further.

- It is here puruSa (the dark matter) invokes the preponderance of Sattva/Entropy in these tamasic/ molecular clouds once again by contracting them and creating stellar nucleosynthesis processes, thus evolving Stars or Aditya, like our Sun.

- During these stellar nucleosynthesis processes, which create huge energies (Rajas), matter is thrown out or spread due to explosions, which become rocks (without any atmosphere) that simply reflect the energy of Sun. These are the Candram.

- Some of these matter that are thrown out are so massive (guru) that they capture a gas envelope around them (varAnakam) and become planets like Earth or Krishnam where biological beings can evolve and survive. Earth is called KrSnam as it is ploughed. KrS means cultivating. krSnam is also used to indicate black because soil of earth is black.

- So, every time when 'tamas' acquires preponderance in a subject of evolution, some observer or some puruSa kick-starts the preponderance of entropy or information content or knowledge once again in that subject of evolution and the cycle of Sattva-Rajas-Tams repeats progressing the evolution. This cyclical model of trigunas progressing the evolution is observed across different domains of the Universe.

Uddalaka says, if in Sun (Aditya) or Moon (Chandra) or anything that is illuminated (vidyuta), Teja (entropy), Apa (energy) and Annam (Mass) are removed, then they won't exist as them. Though all three gunas are present in them in different proportions, the Sun is predominantly Teja, Moon is predominantly Apa and Earth is predominantly Annam.

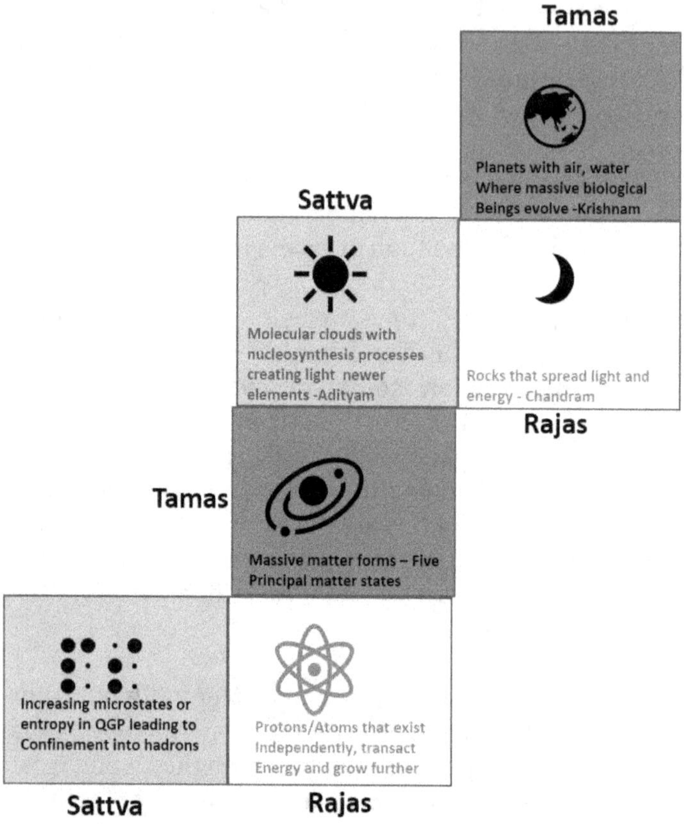

In our Solar system, Sun arises on its own from a solar nebula or molecular clouds, when entropy of molecular clouds increases, due to spacetime curvature impact, even without any addition of external energy, it results in nuclear fusion. This entropy increase is due to dark matter which remains an observer not taking part in evolution of matter. This increasing entropy is the Sattva or *Tejas*. The increased entropy in turn increases the energies which throws out pieces of rocks that are less or more massive. The pieces of rocks that are less massive and do not hold an atmospheric envelope spreads the energy of the sun. They are the 'moon'. Moon is equivalent to Rajas or *Apa* as it spreads the energy of sun to the planets. Moon is called Candra because it spreads or reflects the energy or light of the Sun. The pieces of rocks that are massive enough to hold an atmosphere becomes planets that can sustain biological life. Earth and other such terrestrial planets where more massive forms evolve are the Tamas or *Annam.*

On this Tamasic earth, biological beings evolve again by a cycle of varying gunas of Sattva, Rajas and Tamas. No biological evolution can take place in earth until organic polymer chains form genetic material that can encode other chemicals like proteins. This is again the Sattva or Entropy or Tejas being kick-started in a Tamasic environment. This is done by Catalysts which do not take part in evolution but enable these polymer chains of genetic material to form. These are the virus kind of intermediate beings that hold genetic information but do not metabolize energy. From these beings that hold genetic material arises beings that can metabolize energy on their own. This is Rajas or Apa or Energy acquiring predominance again. These are the bacteria, algae and other micro-organisms that exist independently. From these energy metabolizing beings, arise massive beings that occupy the planet Earth. The plants and animals are in this category. These massive biological beings are the Tamas or Annam or Mass acquiring predominance again.

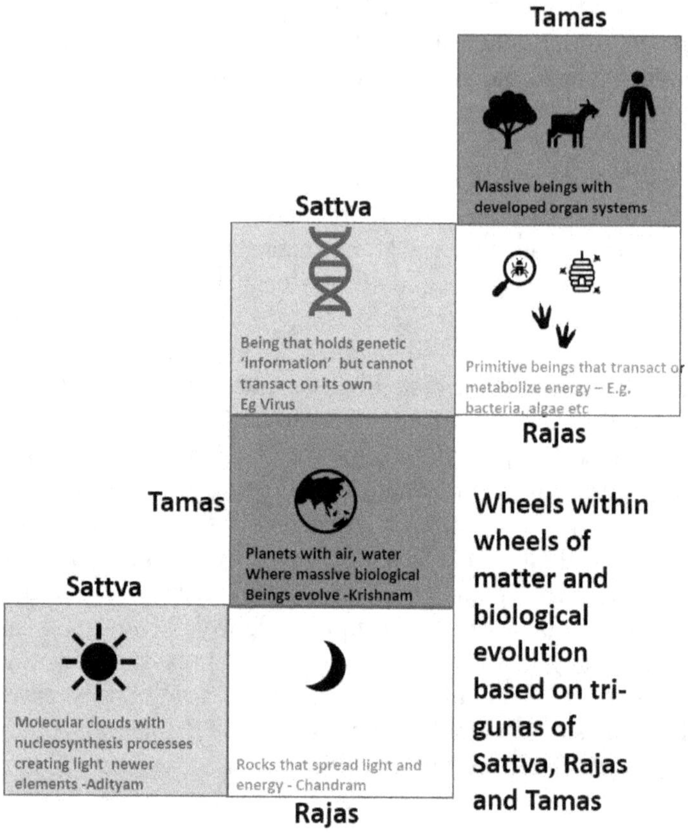

The reproduction pattern of the biological beings also follows the triguna pattern. Uddalaka also cites examples of three types of biological beings. Udbhijam arise on their own. Though it is traditionally interpreted as those that arise from seeds, 'Udbhijam' are those that reproduce by division and hence said to be arising on their own. It's like the entropy of these beings increase rather than their energy metabolizing capability as they only divide. This is the Teja or Sattva. From these beings arise arise those beings that lay eggs or provide seeds externally. Beings that arise from eggs or seeds laid externally are the andajam. Though these are interpreted as 'egg-laid' creatures, these are actually beings that arise from seed or eggs that get distributed and thus spread all over the place. Here because the seeds, eggs, spores etc are spread and arise again in different environments, their energy metabolizing capability continuously evolves. This is the Apa or Rajas, the energy. From these arise the jIvajam, the beings that give birth to their own by sexual reproduction. Because of adaptability and natural selection, these beings grow massive dependent on environment. These are like annam, the food grain that give rise to more food grains or the Tamas.

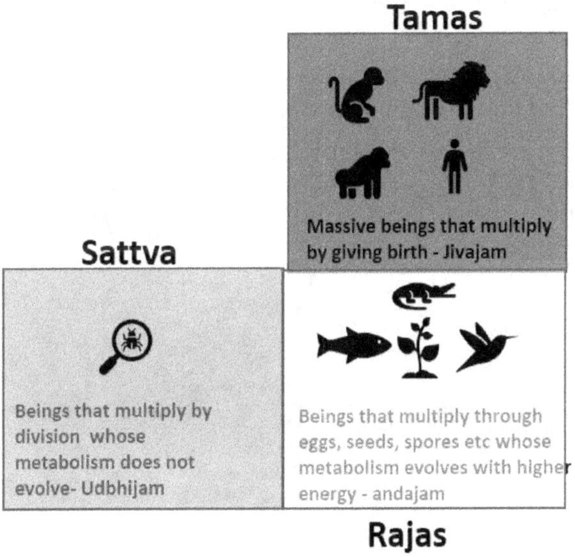

Thus, all known and unknown matter and biological forms and their various characteristics evolve by a varying combination of the trigunas.

Sattva	Entropy	Adityam	Udbhijam
Rajas	Energy	Candram	Andajam
Tamas	Mass	Krishnam	jIvajam

The evolution of thinking human beings also follow the triguna design pattern. The three gunas of Sattva (Teja), Rajas (Apa) and Tamas (Annam) also manifests as Manah, prAna and vAk in human body.

- vAk, the prajna or consciousness is the 'information' or basic signaling that gives awareness. It is the consciousness. It is equivalent to entropy/information content, the Teja or Sattva.
- prAna is the energy providing metabolism that allows beings to build and evolve. Breathing is the first of the metabolisms and hence prAna is often associated only with Breathing. But there are five prAnas and hence called pancha-prAna. This is equivalent to the energy that spreads, which is the Apa.
- Manas, the mind, is thoughts attached to the body and its actions. Thoughts rise one on another like foodgrains (annam) or like massive particles. It is the annam.

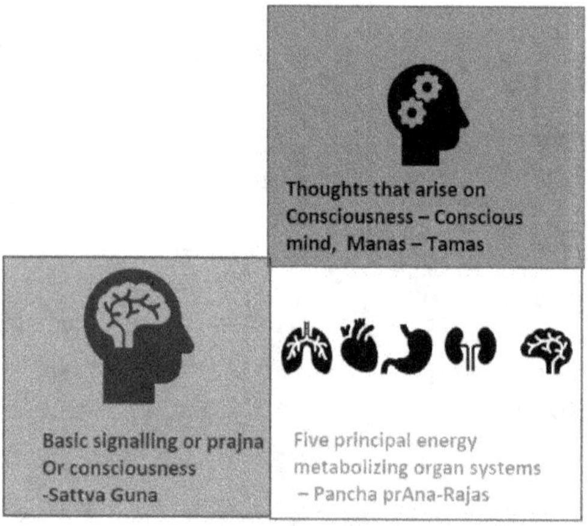

Human Body

38 MANAS – THE CONSCIOUS MIND

Chandogya Upanishad Sloka 6.1 and 6.2 state

dadhnah somya mathya-maanasya yah anima sah oordhvah samudeeshati tat sarpih bhavati.
evam eva khalu somya annasya ashaya-maanasya yah anima sah oordhvah samudeeshati tat manah bhavati.

When the curd is churned (dadhnah mathya-maanasya), the subtlest part (sah anima) detached on top (oordhvah samudeeshati) becomes that butter (tat sarpi bhavati). Same way from the seat of the feelings/emotions (azaya mAnasya), the subtlest part (sah anima), floats on top (oordhvah samudeeshati) that becomes mind (tat manah bhavati).

Azaya means an abode or seat. When this seat of manas (sub-conscious mind) is churned, thoughts arise as the subtlest part, the curd. Manas are the thoughts that arise on top like butter on the curd. Thus, Manas is the conscious mind on top that controls our actions. Hence manas are described as thoughts attached to our body and its actions. These thoughts of conscious mind produce more thoughts, build over one another, as annam produces more annam.

vAk	Consciousness/ Prajna	Fire	Teja
prAna	Breathing/ Organ systems	Water	Apa
Manas	Conscious mind/Action driving thoughts	Food	Annam

The evolution of manas, the thoughts of the conscious mind, in itself follows the triguna design pattern. All evolutionary modifications in our manas, the domain of thoughts result in an increase in knowledge or passion or reduction in knowledge (ignorance) or passion (laziness).

- When diverse types of information are received and processed, knowledge or information content (and hence information processing capacity) increases, it is sattva guna.
- When the same types of thoughts are received and processed, passion increases. It is called rajas guna.
- When less or no information is received, indifference or ignorance increases around that. It is called tamo guna.

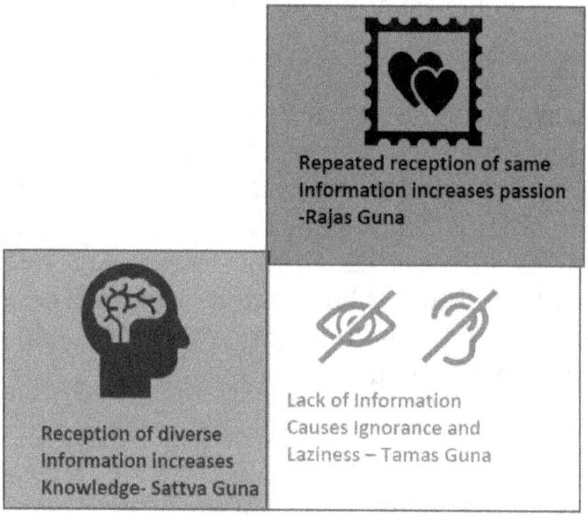

Human thoughts

Thus tri-guna design pattern can be seen in matter, biological beings, thoughts and even in the human civilization. The above design patterns of tri-gunas are a common theme found in multiple scriptures. The tri-gunas of Sattva as increasing entropy, Rajas as increasing energy and Tamas as increasing mass is exhibited all through evolution in different domains. Hence evolution can be seen as cycles within cycles of differentiation caused by the tri-gunas.

39 TAMAS – IMPORTANT MODIFIER

Then Uddalaka explains that the most important modifier or differentiator amongst the Teja, Apa and Annam is the Annam. Annam is equivalent to the manas, the thoughts that control our actions that is the most important differentiator. Annam is equivalent to the Jivajam, the beings that arise from other beings that are the most important differentiators. Annam are the food-grains that build mass of human beings. Annam is the tamas of saMkhya kArika.

Since manas is annam (or food), he asks Svetaketu not to eat food for 15 days and then recite the Rg, Yajur and sAma. But Svetaketu who did not have food for 15 days could not recite it. Uddalaka explains that 'annam' (tamas guna) is the most important modifier for body to sustain itself. The same pattern of tamas guna being the most important modifier can be observed across domains.

- Mass is the most important modifier in the evolution of the Universe amongst energy, entropy and mass. If the amount of mass reduces, Universe structures perish. Even the lightest of massive particles (light particles) bring back all the mass in the Universe as they build and evolve on each other.
- Manas, the thoughts that control our body and its actions, is the most important modifier amongst consciousness, metabolism and thoughts in our body. When manas goes away everything goes away. With even a bit of 'Manas', our thoughts, everything can be brought back.
- Jivajam, the beings that arise from other beings are the most important modifiers amongst all living beings. When they reduce, the overall modification on planet earth reduces. Even a bit of them can modify the planet earth in a big way.

Then Svetaketu eats food and then he is able to recite. Like a small dried leaf rekindles the entire fire, even a small addition of tamas guna can kindle guna modification and further evolution.

Uddalaka explains that like birds tied by a string fly around in all directions but settle down in the original place, manas, the thoughts that control our body and actions, wanders around in all directions but settles down on the prAna, the energy providing metabolism. This design pattern can also be seen across multiple domains. Massive particles (tamas) settle on the energy (rajas) in their bindings to evolve into more massive particles. jIvajam settle on andajam for their energy needs.

Uddalaka then explains that puruSa is said to be in a dream (svapna) always (just witnessing) and not participating. That's the inner nature of puruSa. This is same as what is said in saMkhya kArika. Uddalaka says there are sixteen modifiers (or parts) of which 'puruSa' is the witness or observing one, but that modifies. This is also similar to what is said in saMkhya kArika. Uddalaka says that Apa (the energy that spreads) moves things away from puruSa, bringing into existence, all the Universal matter and beings. This is similar to Sattva (Tejas), the increase in entropy, though kick-starts prakRti's evolution, it is increase in Rajas, the energy that creates independent and new forms that actually moves prakRti away from puruSa. Uddalaka calls all the matter and biological forms of existence as 'Sat' which is the 'reality'.

This Sat is the prakRiti in saMkhya kArika. The three types of modification viz., Tejas, Apa and Annam which are the Sattva, Rajas and Tamas bring about all the evolution into different matter forms and biological beings and even our thoughts as described above.

- In saMkhya kArika, puruSa is said to be crippled and cannot evolve. But it facilities the preponderance of Sattva guna in prakRti. In Chandogya upanishad, puruSa is said to be sleeping (not interacting). It facilitates the 'Tejas' in Sat, the existence.

- In saMkhya kArika, the sattva differentiates prakRti and manifests several forms of prakRti. In Chandogya Upanishad, Tejas is the brilliance that makes something manifest or perceived. Hence Tejas is equated to fire.
- In saMkhya kArika, the Rajas guna acquires preponderance which enables these forms to transact energy or passionately interact with each other. In Chandogya Upanishad, Apa is said to move away from puruSa and equated with water.
- In saMkhya kArika, tamas guna enables the matter forms to become more massive or their structure grows bigger by reproduction. In Chandogya Upanishad, Annam is said to be food grains that product more food grains.
- In the evolution of Classical matter forms, dark matter does not interact. It facilitates the entropy in matter forms. Then arises the ability in these matter forms to transact energy. Then arises the ability to produce larger and larger structures.
- In the evolution of biological beings, catalysts do not interact. But they facilitate the beings with genetic material. Then beings that can metabolize energy or transact energy arise. Then more massive biological beings arise.

40 'ATMAN' IS 'SATYAM'

Uddalaka says

sah yah eshah animaa; aitadaatmyam idam sarvam;
tat satyam, sah aatmaa; TAT TVAM ASI, shvetaketah iti.

The 'Satyam' present subtly in everything is that Atman. That is present in you, Svetaketu.

Across the evolution of different matter forms and beings, there is a subtle property of 'existence' called 'Sat. All of them 'exist'. The forms of matter change, but the existence continues. Like Clay, Gold, Iron and other metals that change shape into different forms but retain the core property of the metal, matter changes its size, shape and form, but retains the core property of 'existence'.

'Sat' is Existence. The property of existence that is present in all matter forms and beings is 'Satyam'. Atman manifests as 'Satyam', the property of existence in all matter forms and beings according to Uddalaka. Because 'Satyam' is the core property of what actually exists, it is equated to 'truth' or 'reality'.

Like the pollens of different flowers merge in honey and honey carries the essence of different flowers, Atman is carried as the core property of existence over and over again, as multiple matter forms interact and give rise to new matter forms and biological beings.

Water from the oceans become rivers and flow back to the ocean once again. But in between rivers are called 'Eastern rivers. 'Western rivers' etc., but rivers themselves do not know what rivers they are, from where they came and where they are going. Likewise, all these creatures do not know that they have come from the existence.

Like the sap of a tree is present in root, shoot, branches and its leaves, the Atman, in the form of core property of existence, is present in all the matter and beings of the Universe.

Like the seed of the banyan tree, which if broken may not reveal any inner parts and appear to have arisen out of nothing, while it has come out of something, Universe might appear to have come out of nothing, but it comes out of 'minute-ness' (animAnAm) and has evolved over a period.

Like the salt dissolved in water makes water salty all through, Atman is dissolved as the core property of existence in all matter and beings. With all above similes and examples,

Svetaketu wants more explanation. So Uddalaka explains further with three analogies.

41 Atman as Observer

A man from gAndhAra nation is kidnapped blindfolded and abandoned in a remote place after removing his blindfold. Now he does not know where he is. So he asks people the way to gAndhAra. They give him directions. Using their directions, he goes back step by step towards the gAndhara nation. He uses enquiry (pricchan), experts advise (panditah) and his own judgement (medhAvi) in going back to his place, facilitated by observers and witnesses of himself.

Thus, in this Universe, the guiding and observing thoughts that facilitate us to do enquiry, use experts advise and our own judgement (arrive at decisions) is the AchArya or the Guru. One who has such an Acharya internally knows the puruSa, as Acharya is the facilitating puruSa. The only delay for getting freed are the steps to be taken to reach the destination and not lack of knowledge of destination. This facilitating Guru, the observing thoughts, is the manifestation of that Atman. Thus, the Atman is present as the property of existence in the form of witnessing thoughts.

42 Atman as Consciousness

A man is sick (and lying) people around him keep asking "Do you recognize me", "Do you recognize me". As long as his consciousness (vAk) and thoughts (manas) are not merged (na vAk manas sampadyate), metabolism/prAna is connected to thoughts (manah prAne), consciousness (tejas/vAk) is connected to metabolism (prAna tejasi), consciousness is connected to the outside elements/devatas (tejah parasyam devataayaam), the person knows them.

In this case the person's consciousness is responding to outside environment, Consciousness drives the metabolism and metabolism drives his thinking. Everything's in order and person recognizes others.

When his thoughts get merged into his consciousness (say in comatose state), even if the thoughts are connected to metabolism, consciousness drives the metabolism, outside elements are connected to consciousness, the person does not know.

In this case everything is in order except that the thoughts of the person have merged into his consciousness (gone into a comatose state), which means he loses his cognition and does not know the people, though he is alive. The person becomes just a bundle of consciousness that cannot respond.

In the previous story, Atman manifests as witnessing thoughts, which is the core property of existence (satyam). In this story Atman manifests as the consciousness, which is the core property of existence (satyam).

43 ATMAN IN ALL OUR ACTIONS

A person is accused of committing a theft. They bring a heated axe for that person to hold. If the person indeed committed that act, that falsehood is done by that person. In that case, that person is connected to falsehood. That person is internally sustained by falsehood. The person catches the axe, burnt and punished. If the person did not commit that act, then truth is done by that person. In that case, that person is connected to truth. That person is internally sustained by truth. The person catches the axe, not burnt and released.

The property of existence, the 'satyam' is present in both people. It is connected to and surrounded by falsehood or truth differentiates their actions. This is like a lie-detector test in those days. Those who are surrounded by falsehood are afraid that heated axe will burn their hands and are afraid to catch it. So, they refuse or make tantrums. Those who are surrounded by truth think that they will not be burnt by the axe and hence they go and catch it. It is similar to promising on lit camphor in the later days. Though Atman is present as the property of existence even in those who lie, cheat or rob, their actions are surrounded by falsehood. Atman remains as the property of existence (satyam) even in them.

Thus, with these three examples Uddalaka explains that Atman manifests in human beings as observing or witnessing thoughts in us that guides, as Consciousness of life and in all our actions surrounded by falsehood or truths.

Uddalaka started with the question "Did your teachers teach you "through which unheard becomes heard, unthought becomes thought, unknown becomes known...?" Then he goes onto explain that Atman is the unthought, unknown and unheard.

He then continued explaining that the matter and biological forms get differentiated due to the trigunas (energy, entropy, mass) and manifest in different forms. This unheard, unknown and unthought Atman becomes the core property of existence in all the matter and biological forms. Uddalaka specifically tells Svetaketu that this property of existence, Satyam is in Svetaketu. Svetaketu does not understand this part. What is this 'Satyam', the core property of existence in Svetaketu...?

Uddalaka explains that Satyam is the detached observer that guides Svetaketu like third parties guide the person who is lost. That Satyam is the consciousness that differentiates life and death in Svetaketu, as it does in any other person. That Satyam is present in all the actions done by Svetaketu surrounded by truths or falsehoods of Svetaketu. Svetaketu now has all questions answered and thanks his father Uddalaka.

The perception of 'I'

Chandogya Upanishad Chapter 6 clearly explains how the Atman is present as the property of existence in all matter and beings and hence is called 'Satyam'. It also talks of three forms of modifications lead to all kinds of matter and beings seen in the Universe. It elaborates on the design pattern in the Universe, as described in saMkhya kArika.

Though we understand that all matter and beings evolved one over another and we all are part of that evolution, we all have a perception of 'I' or 'Us'. We associate this 'I' of ours with that Atman and no other matter and beings in the Universe.

How does that perception arise...?

44 EVOLUTION DUE TO TRI-GUNAS

All the matter and beings possess the three 'gunas'. SamkhyA calls it Sattva, Rajas and Tama. Chandogya Upanishad calls it Teja, Apa and Annam. The imbalance of tri-gunas keeps driving the evolution of the Universe.

Sattva is the differentiation due to entropy or number of microstates of a system. It becomes the ability to interact with/learn from these interactions. Rajas is the internal energy that creates a new entity which can evolve or grow further. It becomes to ability to exist and move as one independent entity. Tamas is the differentiation due to mass. It becomes the ability to become heavier and in turn the inability to move or inability or learn more and more.

All matter and beings have different combinations of the above gunas.

- In Universe's terms, the number of microstates in a system (sattva), the energy of a system (rajas) and the mass of a system (tamas) varies across different structures and forms of matter.
- In biological terms, the intelligence level of a being (sattva), the energy metabolizing characteristics of a being (rajas), mass of a being (tamas) varies across different biological beings.
- In thoughts domain, the information content or knowledge in thoughts due to reception of diverse information (sattva), passion in thoughts caused by reception of the same information again and again (rajas), ignorance caused by lack of information (tamas) varies across different people's thoughts.

45 EXPRESSING AHAMKARA

As the matter evolves from atoms to molecules to compounds to cells, these entities acquire different guna combinations which translates as internal property or ahamkAra that allows them to sustain, build and evolve further. Every entity has its own guna combination. The guna combination results in unique own internal property or ahamkAra of every entity.

For e.g., Sodium has a different guna combination than chlorine, sodium chloride has a different guna combination than sodium and chlorine. Thus, sodium chloride has its own ahamkAra different from the ahamkAra of sodium and chlorine. Hence sodium, chlorine and sodium chloride interact and grow in different ways.

This ahamkAra, the internal property that enables to interact, grow and evolve due to guna combination expressed as whole of the constituent components, differentiates an 'entity' from other entities. This ahamkara rises in all matter forms and beings, in all entities when the entities react or interact.

But the way in which this ahamkAra is expressed is limited in primitive forms of matter and wider in more complex forms of matter. A stone has an ahamkAra, an internal identifiable property or a sense of self or ego. But the way it can express that ahamkAra (internal property) is extremely limited and subdued. Hence it is not perceivable to others. In other words, it is dumb.

A bacterium has an internal identifiable property based on guna combination or ahamkAra which is expressed in more ways than stone. Similarly a plant has more options to express its ahamkAra than bacteria as it is more complex. Animals have much more options to express ahamkAra as they are more complex.

Humans at the top of the evolutionary chain in Planet Earth have a lot more options to express their ahamkAra, their internal identifiable property or ego or 'sense of self' when they interact.

For e.g. our brain or heart or lungs or liver or kidney has their own ahamkAras or internal properties, in their interactions. Together as one functioning being, there is an ahamkAra for us as a person which is the combination of all these aham-kAras. On top of it, human beings have 'manas', the thoughts that control our actions, which can also express its own ahamkAra.

An organization as a whole as a collection of human beings has its own ahamkAra which comes out in its interaction with other organizations. This organization could be a community, business or country. In future when we interact with beings on other planets, this could be the ahamkAra of people of earth.

The difference between a stone and a bacterium and a human being and a society is their ability to express their ahamkAra in different ways. As matter and beings build up more and more, their ability to express ahamkAra becomes more diversified.

For e.g. a stone expresses its ahamkAra (inherent property) when it is made to react with some other elements. It does not have much choice. A bacterium moves around and has a limited choice to interact with other beings and express its ahamkAra. A fairly advanced being has more choices in expressing its ahamkAra with flight or fight responses. A human being can express ahamkAra through thoughts and speech as well. Thus, a stone has the same ahamkAra as a human being. It just expresses that ahamkAra in a smaller number of ways than a human being.

46 AHAMKARA IN THOUGHTS

As Chandogya Upanishad says, Atman is the property of existence (satyam) in all matter and beings. It manifests as the detached observer in us and evolves us. But the capability to express the ahamkAra through thoughts and through speech gives rise to the perception of 'I' in human beings. With the capability to express ahamkAra through thoughts, with this perception of 'I', 'I' is perceived as the 'owner' of the body and its actions.

Because of this 'I' perception, the need to sustain this 'I' for long arises in the thoughts. This leads to desire for longer life or even immortality even at the cost of other beings. The desire for the immortality of 'I' makes us to map this 'I' with that Atman, which is beyond all the existence. From this desire various philosophies arise.

Not this Not this

Atman is present inside all the matter and biological forms as the property of existence as described in Chandogya Upanishad Chapter 6. So how can it be described...?

The manifestation of that Atman is puruSa. PuruSa is the facilitator of evolution. PrakRti is the object of evolution, which is in equilibrium, but whose equilibrium is disturbed by puruSa. PuruSa facilitates the evolution of prakRti, but just stands apart from prakRti and remains just an observer or witness to the evolution of prakRti.

'artha' is the purpose or meaning of the Universe. The purpose or meaning of this Universe is the matter content in the Universe that keeps evolving. Hence evolving prakrti is called 'artha'.

Puruṣasya darshanārthaḥ, kaivalyārthas tathā pradhānasya |
pangvandhavad ubhayor api, saṃyogas tatkṛtaḥ sargaḥ |21|

PuruSa's observation of arthah, standing alone from artha, but acting as the prime-mover, the pradhAna, from the union of this lame and blind, creation happens.

47 The 'Blind' Artha

'Artha', the evolving prakRti does not see the puruSa, as puruSa does not contribute any energy to the evolution of prakRti. In turn, puruSa is crippled as it cannot move or evolve. In this union of puruSa and artha, the Universe's creation progresses, according to saMkhya kArika.

Human beings are part of this 'artha' of Universe which cannot see puruSa, the facilitator of evolution. How does a person who is blind understand something...? The person does it by comparing what is unknown with something known to that person.

Same way when we try to understand the puruSa, we compare it with known forms of 'artha' and say 'not this' 'not this'. Since puruSa is not comparable to any known forms of 'artha', we will end up only saying 'not this' 'not this'. This is Neti (na iti - not this) Neti (na iti - not this).

48 PURUSA IS LIKE AN ESSENCE

This is explained in Brihadaranyaka Upanishad Sloka 4.4.22.1

tasya ha etasya puruṣasya rūpam |

yathā māhārajanaṃ vāsaḥ, yathā pāṇḍvāvikam, yatha indragopaḥ, yatha agnyarciḥ, yathā puṇḍarīkam, yathā sakṛt vidyuttam;

sakṛdvidyutteva ha vā asya śrīrbhavati ya evaṃ veda;

Atha atha ādeśaḥ—neti neti, na hy etasmād iti nety anyat param asti;

atha nāmadheyam—satyasya satyamiti;

prāṇā vai satyam, teṣāmeṣa satyam || 3 ||

This is the form of the 'puruSa'. Like perfume (vAsa) of saffron (maharajana), Like jaundice that makes one pale/yellow from inside (pAndvavikam), like a firefly (Indragopa), like a flame of fire that firefly emits (agni arci), like a white lotus (pundarikam), like the radiance (vidyuttam) it emits every day (sakRt). Who knows it (ya evam veda) becomes wealthy/radiant (zrir bhavati) of that radiance?

The saffron's perfume comes from inside. The jaundice that makes one pale comes from inside. The firefly's flame of light comes from its inside. The lotus that closes every night and blooms everyday with radiance comes from its inside. PuruSa seems like all these. One who knows/understands this is endowed with that radiance.

Now therefore (atha atha) according to this instruction (adeza) 'not this not this', nothing else is superior to (na anyat param asti) not this (na iti). Therefore (atha) it is given the name (nAma dheyam) the truth of the truth (satyasya satyam). prAna is Satyam, all these are satyam, thus the three are the brAhman (vAk, prAna, manas).

puruSa seems to be like something deep inside, which cannot be figured out. So, it is best described by Not this, Not this. Hence Not this, not this is given the name of truth of truth or the ultimate truth.

Atman is indeed present inside all the matter and biological forms as the property of existence as described in Chandogya Upanishad Chapter 6. But it cannot be described by the matter or biological forms. Hence Neti, Neti.

Law and Order

If samkhya is a design pattern, puruSa is the facilitator of evolution as a manifestation of Atman, manas are the thoughts attached to the body, Atman manifests as observing or witnessing thoughts, all modifications in Universe can be seen as three forms of modification, then the question that comes up is, is Universe based on 'determinism' or 'randomness'..?

Some people say "Universe is predetermined. There is a purpose to everything. When we look back, we will get to know the purpose. For e.g., look at how earth is placed at the exact distance from the sun that allowed biological beings to evolve" etc. to justify the determinism in Universe.

Some people say "Universe was born out of a Chaos. It is an accident. It could have happened in anyway, but just that it happened in this way and biological evolutions have proceeded".

49 UNIVERSE SEEMS RANDOM

Randomness is lack of a pattern, lack of predictability in events. It is about future prediction ability and not reasoning about the past.

It is absolutely true that we can always connect the dots looking backwards. Since it is our thoughts that connects the dots, we can connect the dots looking backwards in multiple ways. That does not show proof of determinism or lack of randomness.

For one earth that evolved at the exact distance from the Sun, there could have been millions of rocks that did not make it in this way. In millions of galaxies around the universe, so many stars and planets exist and they keep getting destroyed and created. These do not show proof of determinism or lack of randomness.

50 Dharma and Rta

What we all know is Universe is governed by fixed laws. Dharma are the 'natural' set of laws that govern the Universe or any subset of it. This dharma, the natural laws manifests in Quantum domain, Classical domain, biological domain and thoughts domain. Dharma, the natural laws of the Universe seems fixed. A society of human beings align their 'artificial' laws of living to the natural laws of the Universe and call it their dharma.

This dharma, the fixed laws of Universe, brings in Rta or Rhythm or predictability or Order to the Universe and its subsets. Rta simply is the English rhythm. It is possible that word rhythm came from the word Rta. Rta and Rhythm simply means predictability or a predictable set of repeating events. This predictability or rhythm brings in an 'Order' to the Universe. Hence Rta, the predictability is translated as Order also. Seasons are called Rta because there is a predictability to the seasons.

For example, in a human society, the laws of the society, a Rta or predictability or Order is built in the society. This dharma and Rta are the 'law and order' of the society.

This Rta or Rhythm or predictability or order based on dharma, the laws of the Universe is a hall-mark of the current Universe we observe. This Rta or Rhythm manifests as an order of one building up over another. Even if human reasoning connecting the dots back is removed, it is clear that the Universe has evolved over a set of laws on which predictable set of events occur and an order is built.

Today science can predict the future course of Sun, earth and even Universe to certain set of possibilities with various degrees of certainty. We know earth is going to die, the sun is going to die and possibly the Universe as we know will also be extinct. Hence there is a Rta or order in the Universe's evolution.

In other words, those that follow the dharma, the laws of the Universe and evolve in a predictable order only have sustained. Those that do not follow the dharma or evolve with an order, perish back to where they evolved from.

51 RTA – THE ORDER

From the way quarks manifest in dirac-fermi fields to them building up the nucleus, formation of atoms, elements, compounds, galaxies, stars, planets and life, there is an order in the Universe in which one builds up over another and following of certain set of laws. As more complex beings form, more and more complex laws operate them, as several laws that operate each of these components intermingle and crisscross each other.

Look at a contemporary example. A mechanical clock can be repaired even now. But when an LCD TV does not work, one technician will change the entire motherboard, another one will change some IC chip in the motherboard blindly and try fixing it, rather than understand what's really not working. It is tough for them to find what is not working as it has a lot of complex laws operating them. When laws are complex, it is not possible to determine what exactly is not working.

Our life is much more complex. We are all inter-related socially and even physically. Our life is not guided by our thoughts alone or our internal properties alone. Our life is influenced by the thoughts and actions of other people, events of the Universe, laws of the society we live in and far off from us and even the laws of the universe. When the laws are complex, it is not possible to determine or predict the way we progress in our life.

When it becomes too complex to understand, the determinism or predictability is lost. When predictability is lost or when are unable to determine due to complexity, we call it randomness, which is what happens in the Classical domain.

In the quantum domain and in our consciousness domain on which our thoughts evolve, randomness is in their very manifestation. Randomness is the very nature of these two domains. But that randomness does not lead to chaos.

In Quantum domain, laws such as observer effect causes wave function collapse due to observation or measurement. This causes the collapse of probabilistic wave nature of particle to deterministic particle nature. This is demonstrated in the double-slit experiment, where observation or measurement by device collapses the wave nature into particle nature. However, the observer in quantum domain is another quantum particle and not human consciousness.

Thus, the dharma or the laws of the Universe brings up predictability or order in the quantum domain.

52 Observation brings Rta

This observer effect is applicable in several domains including our thoughts. While thoughts may be running helter-skelter, observation of thoughts by the 'Self' (detached witnessing thoughts) collapses the randomness, establishes a predictability or order in the way we think. In fact, it applies to even our action. When our actions are observed we behave differently, making the way we behave more deterministic or orderly.

In any organization (be it business or social) we observe others and are observed by others. This brings in a predictability or order or Rta to the actions driven by our thoughts in that organization. Observation that brings up the reality is the law of the Universe in all domains. It is the dharma. This law brings in a predictability or Order in the Quantum and consciousness domains too which have randomness in their very manifestation.

When it comes to individuals, the best of the observer that can bring predictability and order to our thoughts is our own observing self. Our thoughts that are detached from our own actions, that witness and observe us and others with equanimity is the best observer that brings predictability or Rta or order to our thoughts. This detached witness in us is the manifestation of that Atman, is the observer that drives evolution. This is our manas-sAkshi, the witness to our manas. This is the design pattern across different domains of our Universe.

References

1. The Brihadharanyaka Upanishad - Translated by Swami Madhavananda 1950
2. Ashtavakra Gita - Sanskrit text and Translation - John Henry Richards
3. Sri daksinamurthy stotram - By Adi Shankaracharya
4. Long version of Gurugita- From sanskritdocuments.org
5. saMkhya kArika of Isvara Krsna with Tattva Kaumudi of Vacaspati Misra
6. Chandogya Upanishad Chapter 6 from Reflections by Swami Gurubhaktananda of Chinmaya International foundation
7. Vishnu Sahasra nAma Stotram- S.V. Radhakrishna Sastry - 1986
8. Inspiring Palita perspectives - Chapter - Samkhya - The design pattern

www.ingramcontent.com/pod-product-compliance
Lightning Source LLC
Chambersburg PA
CBHW070444220526
45466CB00004B/1768
* 9 7 8 1 6 7 1 3 1 1 8 5 5 *